Plate 1. Outdoor beds showing the use of bedding plants to provide large splashes of colour.

Plate 2. Window boxes are particularly suited to the use of annuals.

Plate 3. Bedding plants make useful accents in a mixed border.
Plate 4. An example of seed coating on bedding plant species.
Plate 5. Leaf discoloration on foliage of *Pelargonium* × *hortorum* due to chlormequat application.

CROP PRODUCTION SCIENCE IN HORTICULTURE series

Series Editors: Jeff Atherton, Senior Lecturer in Horticulture, University of Nottingham, and Alun Rees, Horticultural Consultant and Editor, *Journal of Horticultural Science*

This series examines economically important horticultural crops selected from the major production systems in temperate, subtropical and tropical climatic areas. Systems represented range from open field and plantation sites to protected plastic and glass houses, growing rooms and laboratories. Emphasis is placed on the scientific principles underlying crop production practices rather than on providing empirical recipes for uncritical acceptance. Scientific understanding provides the key to both reasoned choice of practice and the solution of future problems.

Students and staff at universities and colleges throughout the world involved in courses in horticulture, as well as in agriculture, plant science, food science and applied biology at degree, diploma or certificate level will welcome this series as a succinct and readable source of information. The books will also be invaluable to progressive growers, advisers and end-product users requiring an authoritative, but brief, scientific introduction to particular crops or systems. Keen gardeners wishing to understand the scientific basis of recommended practices will also find the series very useful.

The authors are all internationally renowned experts with extensive experience of their subjects. Each volume follows a common format covering all aspects of production, from background physiology and breeding, to propagation and planting, through husbandry and crop protection, to harvesting, handling and storage. Selective references are included to direct the reader to further information on specific topics.

Titles Available:
1. **Ornamental Bulbs, Corms and Tubers** A. R. Rees
2. **Citrus** F. S. Davies and L. G. Albrigo
3. **Onions and Other Vegetable Alliums** J. L. Brewster
4. **Ornamental Bedding Plants** A. M. Armitage

Titles in Preparation:
Cucurbits R. W. Robinson
Tomatoes J. Atherton
Carrots and Related Vegetable Umbelliferae V. E. Rubatzky, P. W. Simon and C. F. Quiros
Coffee, Cocoa and Tea K. C. Willson
Bananas and Plantains J. C. Robinson
Horticultural Food Legumes G. D. Hill and J. Smartt
Temperate Zone Herbs L.E. Craker, J.E. Simon and B. Galambosi

ORNAMENTAL BEDDING PLANTS

Allan M. Armitage

Professor,
Department of Horticulture,
University of Georgia,
USA

CAB INTERNATIONAL

CAB INTERNATIONAL
Wallingford
Oxon OX10 8DE
UK

Tel: Wallingford (0491) 832111
Telex: 847964 (COMAGG G)
Telecom Gold/Dialcom: 84: CAU001
Fax: (0491) 833508

A catalogue entry for this book is available from the British
Library.

ISBN 0 85198 901 2

Typeset by Solidus (Bristol) Ltd
Printed and bound in Great Britain by
Biddles Ltd, Guildford and King's Lynn

CONTENTS

1

INTRODUCTION

What is a Bedding Plant?

The definition of a bedding plant is constantly changing. Historically, bedding plants were those plants used for 'bedding out', that is, to provide colour in garden beds. Today, the term has been expanded by some writers to embrace any herbaceous plant started under controlled environmental conditions and sold to home gardeners and landscapers for outdoor use (Carlson and Johnson, 1985). Such a definition could include ornamental annual and biennial plants, herbs, ground covers, perennials, vegetables (tomato, pepper), small fruits (strawberry, potted grapes) and even some woody plants which die to the ground in the autumn (*Buddleia*). What plants are included in the broad definition of bedding plants also depends on the country in which they are grown. In the United Kingdom, for example, traditional fruit, vegetable and woody plants are not included. Plants may be offered in various containers, ranging from small 'cell packs' to 10–12 cm diameter pots. Although often grown in pots, bedding plants differ from greenhouse 'pot plants' in that pot plants are sold for indoor use, such as gift plants, brightening up the home or bringing fragrance inside. Pot plants include such well-known species as poinsettia, chrysanthemum, Easter lily and kalanchoe.

Bedding plants are a loose aggregation of many species of plants, some of which have multiple uses. For example, species of primrose (*P.* × *polyantha*, *P. malacoides*) are popular pot plants for the florist and gift market as well as important bedding plants in temperate countries. In this book, ornamental bedding plants include tender herbaceous ornamental annuals and biennials only and do not include perennials or species used for fruit or vegetable production.

Dozens of annuals (biennials) are marketed as bedding plants (see Tables 1.3 and 1.4 below) and to grow all species invites major difficulties in scheduling and quality control. Approximately ten species are sufficiently important to be grown by most producers, while others may be found in

1

specialty and niche grower operations. Numerous production and marketing difficulties may be sustained in the production of such a diverse group of species; however, no other flower commodity has been as stable in its growth rate and customer demand. In the United States and Canada, sales have increased by over 15% per annum since the mid-1970s (Nelson, 1991), although sales have stabilized in recent years. In Northern Europe, where use of bedding plants lagged behind other traditional garden and greenhouse crops, rates of sales in bedding plants are increasing every year.

Problems of Production

The production of such a diverse group of species presents numerous problems. The assortment of species demands a number of diverse greenhouse environments for optimum plant quality, a demand difficult to obtain under one holding. Since spring demand is by far the strongest, production occurs during the cold, dull days of winter. Environmental inputs of heat and light must be provided to force plants into flower, and such additions are expensive. Therefore, the greenhouse operator must constantly strive to finish the product over a short time period while maintaining plant and flower quality. Greenhouse operators must juggle the environmental needs for the various species with the ability of their facilities to provide such environments. Areas for germination, growing-on and finishing must be provided for different times during the production season. Scheduling plants for market requires intimate knowledge of the environmental needs of each species, from seed germination to leaf and flower initiation. In some countries, plants must be in flower to be sold. This means that scheduling must become even more precise due to the rejection of green plants in the market place. Where plants may be sold green, scheduling is less demanding.

Scheduling is made more complex when one realizes that the vast majority of bedding plants will be sold over four to six weekends in the spring season. Some countries have a more extended sales period, but regardless of locale, spring is the busiest time. If inclement weather occurs on those days, consumers stay home, plants cannot be moved to retail outlets and they back up in the greenhouse. The profitability of bedding plants is often highly dependent on spring weather; a fickle mistress at best. When it rains on three to four successive spring weekends, and the vinca leaves are yellow, the geranium flowers are ripe with botrytis and the marigolds and dahlias are falling apart because they can't be moved, tomato production sounds rather inviting to the bedding plant grower.

Important Species and Production Areas

The bedding plant market has become increasingly important in Northern European countries as well as in North America. Table 1.1 lists the six most important species in the United Kingdom (Miles, personal communication), France, The Netherlands (EuroFloratech, 1991) and the United States (Behe and Beckett, 1993) in 1991. Production area in 1991 devoted to bedding plants in West Germany was 629 hectares, including outdoor production, while greenhouse production alone was 445 hectares in France and 120 in the Netherlands (EuroFloratech, 1991).

Table 1.1. The top six bedding plants in the UK, France, The Netherlands and the United States, in order of volume sold. Scientific names are listed below the table.

United Kingdom	France	Netherlands	United States
Pansy	Geranium	Geranium	Impatiens
Geranium	Begonia	Pansy	Petunia
Marigold	Pansy	Impatiens	Geranium
Impatiens	Petunia	Marigold	Vinca
Petunia	Impatiens	Begonia	Pansy
Lobelia	Carnation	Lobelia	Begonia

Pansy (*Viola* × *wittrockiana*); Geranium (*Pelargonium* × *hortorum*); Impatiens (*Impatiens* × *hybrida*); Begonia (*Begonia semperflorens*-cultorum); Petunia (*Petunia* × *hybrida*); Marigold (*Tagetes* spp.); Vinca (*Catharanthus roseus*); Lobelia (*Lobelia erinis*); Carnation (*Dianthus* spp.)

Autumn Sales

Because of the dependence on the spring for bedding plant sales, a good deal of marketing and promotion has been conducted in the last decade to promote autumn for planting bedding plants. The advent of better cultivars of cold hardy annuals such as pansies and ornamental kale has resulted in autumn becoming a more important sales season, particularly in areas of mild winters. In tropical areas (e.g. Florida, USA), the same species sold in the spring in temperate areas are sold in the autumn for winter and early spring flowering. In temperate areas, however, species with greater cold hardiness are used for autumn sales, the most important being *Viola* × *wittrockiana* (pansy), but other half hardy annuals are also offered (Table 1.2). The autumn market is not as

Table 1.2. Important bedding plant species for autumn sales.

Crop	Genus and species
California poppy	*Eschscholzia californica*
Fairy primrose	*Primula malacoides*
Forget-me-not	*Myosotis sylvatica*
Iceland poppy	*Papaver nudicaule*
Ornamental kale	*Brassica oleracea*
Pansy	*Viola × wittrockiana*
Pinks	*Dianthus* hybrids
Polyantha primrose	*Primula × polyantha*
Snapdragon	*Antirrhinum majus*
Wallflower	*Cheiranthus chieri*

strong as the spring market, particularly in areas of severe winters; however, it has become more and more important over the last 10 years. With additional cold hardiness and novel flower colours spilling out of plant breeders' laboratories, autumn sales will continue to expand.

Sales Outlets

Bedding plants are sold from numerous outlets, and some types are more important than others, depending on country. In Europe, bedding plants are usually purchased from garden centres and small local vendors, while recent research in the United States showed that garden centres accounted for 46% of sales, discount and chain stores, 24%, supermarkets, 6%, florist shops and street vendors, 6% and mail order catalogues, 8% (Fossler, 1993). Various other forms of retail outlets accounted for the remaining 8%.

OCCURRENCE BY FAMILY

Bedding plants are represented by families of dicotyledonous plants, the vast majority being tender annuals outdoors. No monocotyledons are represented and, with the exception of *Dahlia* and *Begonia* species, no genera possess storage roots. Asteraceae contains the most genera of important bedding plants; however, many plant families may be found in a typical commercial greenhouse operation. Each genus may be represented by one or two species. Table 1.3 provides a list of families with examples of important genera found in the bedding plant trade.

Table 1.3. Plant families and some representatives of bedding plants.

Family	Example	Family	Example
Acanthaceae	*Thunbergia*	Caryophyllaceae	*Dianthus*
Amaranthaceae	*Amaranthus*	Geraniaceae	*Pelargonium*
	Celosia	Lamiaceae	*Coleus*
	Gomphrena		*Salvia*
Apocynaceae	*Catharanthus*		
		Lythraceae	*Cuphea*
Asteraceae	*Ageratum*	Malvaceae	*Abelmoschus*
	Aster		
	Calendula	Nolanaceae	*Nolana*
	Chrysanthemum	Papaveraceae	*Eschscholzia*
	Cosmos		*Papaver*
	Dahlia		
	Dyssodia	Plumbaginaceae	*Limonium*
	Gazania	Polemoniaceae	*Phlox*
	Gerbera		
	Melampodium	Portulacaceae	*Portulaca*
	Sanvitalia	Primulaceae	*Primula*
	Senecio		
	Tagetes	Saxifragaceae	*Saxifraga*
	Zinnia	Schropulariaceae	*Antirrhinum*
Balsaminaceae	*Impatiens*		*Mimulus*
Begoniaceae	*Begonia*		*Nemesia*
			Torenia
Boraginaceae	*Myosotis*	Solanaceae	*Capsicum*
Brassicaceae	*Brassica*		*Petunia*
	Campanula		*Nicotiana*
	Cheiranthus		*Nierembergia*
	Matthiola		
		Verbenaceae	*Verbena*
Campanulaceae	*Lobelia*	Violaceae	*Viola*
Capparidaceae	*Cleome*		

HORTICULTURAL USES

The demand for bedding plants has consistently risen over the past 15 years. The market for bedding plants can be divided into two main segments. The first is the consumer market, catering to the individual homeowner or renter. This market is supplied by retail outlets such as chain stores and garden centres. The retail outlets have made significant investments in facilities and the number of outlets have mushroomed. Spring sales of bedding plants are a primary source of revenue for the majority of outlets, and sales of other plant materials such as cut flowers and pot plants pale in comparison. Important secondary sales include fertilizer, chemicals and tools to complement bedding plants in the garden. The consistent increase in the consumer market is fueled by greater awareness of gardening in many countries and greater availability of plant material. Outlets such as hardware stores, petrol stations, and large discount stores sell bedding plants during the spring, some of which do more harm than good to the bedding plant market (see Postproduction, Chapter 4); however, the fact that plants are more accessible stimulates the demand. The need and desire for persistent colourful plants in the garden or window box continues to grow, and their use in the landscape continues to increase.

The second major market caters to the landscaper and landscape maintenance industries. These groups work with designers and landscape architects to provide seasonal colour for homeowners and commercial developments. Corporate headquarters, retail shops, airports and other public areas demand large blocks of sustainable colour, and landscapers have increased their use of annuals (and other plants as well) due to improvement of plant habit and garden performance of many species of bedding plants. Plants used in such areas are viewed by many people and are often seen from the road as people drive by. Breeders have recognized that some plants must be more vigorous and have developed certain cultivars for the landscape trade. In the United States, such cultivars have been nicknamed '55-mph plants'.

While the interest in gardening and plant availability have increased, so has the quality and diversity of plant material. Plant breeders and seed companies have consistently introduced better cultivars and new species every year. Greenhouse performance has been enhanced through cultivars with better seed germination and accelerated flowering. As greenhouse performance improves, growers are more likely to grow certain groups of plants, thereby filling the market with better quality plants. As cultivars are bred with more compact growth, more persistent flowering and better weatherability, their garden performance is also enhanced, thereby ensuring continued improvement.

Planting Situations

Bedding plants can be planted to provide colour throughout the season. The possibilities include large drifts of colour in outdoor beds (Plate 1) as well as splashes of colour in window boxes and planters (Plate 2). They may also be used to complement small shrubs and perennials and supplement the garden border (Plate 3). In most temperate zones, bedding plants for summer colour are placed in the garden after the last frost date and removed after the first frost. In areas of relatively mild winters, however, plants may also be set

Table 1.4. Annuals for special uses (after Procter, 1992).

Annuals for shady areas (after Springer, 1992)

Begonia semperflorens-cultorum	Fibrous begonia
Begonia tuberhybrida	Tuberous begonia
Coleus alabamensis	Alabama coleus
Coleus × *hybrida*	Coleus
Hypoestes phyllostachya	Polka dot plant
Impatiens × *hybrida*	Impatiens
Lobelia erinus	Lobelia
Lunaria annua	Honesty
Meconopsis cambrica	Welsh poppy
Myosotis sylvatica	Forget-me-not
Nicotiana sylvestris	Woodland tobacco
Torenia fournieri	Wishbone plant

Plants grown mainly for foliage (after Overy, 1992)

Acalypa wilkesiana	Acalypa
Amaranthus tricolor	Joseph's coat
Atriplex hortensis	Red orache
Coleus × *hybridus*	Coleus
Dolichos lablab	Hyacinth bean
Euphorbia marginata	Snow-on-the-mountain
Hypoestes phyllostachya	Polka dot plant
Iresine herbstii	Bloodleaf
Kochia scoparia	Burning bush
Perilla frutescens	Chinese basil
Ricinus communis	Castor bean
Senecio cineraria	Dusty miller

Annuals useful for cut flowers (after Buchanan, 1992)

Ammi majus	False Queen Anne's lace
Antirrhinum majus	Snapdragon
Callistephus chinensis	China aster
Celosia cristata	Cockscomb
Centaurea cyanus	Bachelor's buttons
Centaurea moschata	Sweet sultan
Consolida ambigua	Larkspur
Cosmos bipinnatus	Cosmos
Gomphrena globosa	Globe amaranth
Gypsophila elegans	Annual baby's breath
Limonium sinuata	Annual statice
Moluccella laevis	Bells-of-Ireland
Helichrysum bracteatum	Strawflower
Scabiosa atropurpurea	Pincushion flower
Tagetes erecta	African marigold
Zinnia elegans	Zinnia

Table 1.4. *continued*

Annuals for dry areas

Brachycome iberidifolia	Swan river daisy
Catharanthus roseus	Madagascar periwinkle
Centaurea cyanus	Bachelor's buttons
Cleome hassleriana	Spider flower
Coreopsis tinctoria	Tickseed
Cosmos bipinnatus	Cosmos
Cuphea ignea	Firecracker plant
Dimorphotheca sp.	Cape marigold
Dyssodia tenuiloba	Dahlberg daisy
Eschscholzia californica	California poppy
Euphorbia marginata	Snow-on-the-mountain
Eustoma grandiflora	Prairie gentian
Gaillardia pulchella	Annual Indian blanket
Gypsophila elegans	Annual baby's breath
Helichrysum bracteatum	Strawflower
Kochia scoparia	Burning bush
Limonium sinuatum	Annual statice
Nolana paradoxa	Chilean bellflower
Perilla frutescens	Perilla
Portulaca grandiflora	Moss rose
Rudbeckia hirta	Black-eyed Susan
Senecio cineraria	Dusty miller
Verbena × *hybrida*	Annual verbena

out in the autumn for winter and early spring colour.

Bedding plants may be selected for many different areas, such as wet, shady or full sun areas. Additional uses include species for cut flowers, fragrance or fine foliage. Table 1.4 provides a listing of annuals for certain special uses; however, regional specialists should be consulted for selection of individual cultivars.

TEMPERATURE TOLERANCES OF ANNUALS

Cold Hardiness

Most species of annuals produced for garden use will continue to flower and grow year-round under greenhouse conditions. However, the majority of annuals discussed in this book are meant for seasonal display only, and die due to cold winter temperatures. Many species are tropical in origin and die at first frost, others are moderately cold-tolerant and may tolerate 1–3°C below freezing before succumbing. These species often look best during the cool spring and autumn seasons. Some species tolerate even colder temperatures and may overwinter in climates with moderate winters. Such species are often planted in those areas in the autumn for early spring flowering. Autumn planted pansies are a major bedding plant in the southern United States and in some European countries. Cold tolerance is significantly affected by autumn temperature and water content in the tissue. Plants are more able to withstand some frost when temperatures in the autumn fall gradually, allowing some hardening off of the plant tissue. Warm temperatures in late autumn, followed by a light frost (-1 to $-2°C$) often severely damage or kill plants which are otherwise moderately cold tolerant. Ornamental kale and cabbage (*Brassica oleracea*) can survive temperatures as low as $-10°C$ if the plant is sufficiently hardened off. On the other hand, I have seen cabbage soup at $-2°C$ when it was preceded with summer-like temperatures simply because plants were not hardened off.

Table 1.5. Relative cold hardiness of annuals. The table assumes some gradual cooling prior to frost.

Least cold tolerant	Moderately cold tolerant	Most cold tolerant
Abelmoschus	Ageratum	Antirrhinum
Amaranthus	Aster	Brassica
Begonia	Chrysanthemum	Calendula
Browallia	Dahlia	Chrysanthemum
Capsicum	Dyssodia	Dianthus
Catharanthus	Gazania	Matthiola
Celosia	Gerbera	Primula
Cleome	Limonium	Viola
Coleus	Lobelia	
Cosmos	Lobularia	
Cuphea	Mesembryanthemum	
Eustoma	Mimulus	
Gomphrena	Nicotiana	
Hypoestes	Nierembergia	
Impatiens	Petunia	
Melampodium	Phlox	
Nemesia	Torenia	
Nolana	Verbena	
Pelargonium		
Portulaca		
Salvia		
Tagetes		
Thunbergia		
Zinnia		

Tolerance of cold is also enhanced when plants are turgid when the first frost occurs. Even with insufficient hardening, plants are more tolerant of cold temperatures than when they are well watered and protected from desiccating wind. Table 1.5 provides **relative** cold tolerance for annual genera. Some species within the genus may react differently. For example, *Primula acaulis* is more cold-tolerant than *P. malacoides*. Similarly *Salvia farinacea* is significantly more cold-tolerant than *S. splendens*, although both are grown as annuals in many areas.

Heat Tolerance

Tolerance of warm temperatures is important when plants are used in areas of hot summers. Some species are able to tolerate hot temperatures and continue to flourish, others perform better in moderate summers whereas a few species appear intolerant of extended periods of hot weather. The

Table 1.6. Relative heat tolerance of annuals. The inability to tolerate high temperatures is made worse when humidity is also high.

Least heat tolerant	Moderately heat tolerant	Most heat tolerant
Antirrhinum	Ageratum	Abelmoschus
Aster	Amaranthus	Begonia
Brassica	Capsicum	Catharanthus
Browallia	Celosia	Cleome
Calendula	Chrysanthemum	Dyssodia
Dianthus	Cosmos	Gomphrena
Lobelia	Cuphea	Hypoestes
Matthiola	Dahlia	Impatiens
Mimulus	Eustoma	Limonium
Nolana	Gazania	Melampodium
Phlox	Gerbera	Nicotiana
Primula	Lobularia	Zinnia
Verbena	Mesembryanthemum	
Viola	Nemesia	
	Nierembergia	
	Pelargonium	
	Portulaca	
	Salvia	
	Sanvitalia	
	Tagetes	
	Thunbergia	
	Torenia	

discussion of heat tolerance is fraught with grey areas. For example, plants are more tolerant of hot days if night temperatures decrease 6–10°C below day temperatures. For most plants, respiration occurs twice as rapidly for each 10°C rise in temperature. When night temperatures remain high, respiration, the process of using carbohydrates, continues unabated. The process of producing carbohydrates for plant tissue, photosynthesis, however, stops due to darkness. Under extended periods of high night temperatures, respiration can deplete stored carbohydrates. Under lower night temperatures, plant respiration remains in balance with photosynthetic gains made during the day and plants are therefore more able to cope with high temperatures.

The other common environmental parameter which affects plant heat tolerance is the amount of moisture in the air, usually reported as relative humidity. High moisture content results in lush growth, open stomata and makes the plant more susceptible to numerous fungal diseases. In areas of hot temperatures, plants may already be stressed and the additional susceptibility to pathogens may cause severe damage or death. Therefore, the combination of high humidity and high temperature is more debilitating to plant performance than the same heat combined with low humidity.

Table 1.6 provides **relative** heat tolerance of annual species of various genera. The various categories also reflect intolerance to a lack of day/night temperature differences and high relative humidity.

BREEDING AND SELECTION

The impressive growth of the bedding plant industry has gone hand in hand with the development of new and improved cultivars. In general, the major contribution of the flower breeder is to develop distinct varieties, uniform in many respects, with a high seed quality, high levels of resistances, a predictable cropping period, stress tolerance, wide adaptability and a low requirement for chemical inputs (van Kester, 1990).

One of the more important achievements has been the introduction of F_1 hybrids in many species of bedding plants. In producing F_1 hybrids, the breeder maintains numerous inbred lines of each species. Inbred lines are the result of enforced self-pollination of individual plants through consecutive generations. In many species, inbreeding causes an increase in uniformity but may be accompanied by a decline in vigour and size (inbreeding depression). When two or more inbred lines are crossed, F_1 hybrid lines result. The population is vigorous, heterozygous and uniform, and visually exhibits greater vigour than the parents, known as hybrid vigour (Hartmann *et al.*, 1990). Mixtures may also be offered which are combinations of separate lines similar in characteristics except for one trait (e.g. flower colour). In the commercial trade, separate colours of these mixtures are referred to as a series (e.g. Primetime series of petunias, Impulse series of impatiens, etc.). In some crops (petunia, pansy, snapdragon), F_2 hybrids have also been developed. They are the result of standard open-pollinated or self-pollinated F_1 plants and are relatively uniform, compared to standard open-pollinated cultivars. The main advantage to F_2 seed is that it is less costly to produce than F_1 seed, yet is still reasonably uniform and vigorous. The use of F_1 seed, however, far outweighs the use of F_2 seed in commercial bedding plant production. Almost all F_1 hybrids have been developed since 1950 (Craig and Laughner, 1985), and they continue to be developed today. F_1 hybrid cultivars have been developed in *Ageratum, Antirrhinum, Begonia, Calceolaria, Capsicum, Calendula, Dianthus, Eustoma, Gazania, Gerbera, Impatiens, Nicotiana, Pelargonium, Petunia, Portulaca, Salvia, Tagetes, Viola* and *Zinnia*. Cultivars of a few other species are available through asexual techniques (*Coleus, Gerbera,* New Guinea impatiens, *Pelargonium*).

Breeding is both an art and a science. The art includes selection of improved parents and offspring based on horticultural characteristics. One of the important skills of the flower breeder is his/her ability to evaluate the horticultural attributes of a potential cultivar. With literally hundreds of crosses from which to choose, they must display skilful ruthlessness. The breeder must also scientifically evaluate inheritance patterns of various characteristics and chromosomal variations. The disciplines of plant physiology, morphology, anatomy and taxonomy are the traditional tools of the flower breeder. Gene transfer, micropropagation, male sterility techniques and restriction fragment length polymorphism (RFLP) procedures have recently been developed and should assist breeders to develop additional cultivars (Craig, 1990). For example, anthocyanin biosynthesis was enhanced through the introduction of a maize gene to petunia, resulting in red flowers containing pelargonidin (Meyer *et al.*, 1987), while flower colour patterns in petunia and *Nicotiana tabacum* were modified through gene transfer (Krol *et al.*, 1988). Although no genetically engineered cultivars have yet been introduced, they will without doubt appear in the next decade.

A flower breeder often develops a product profile to generate better cultivars. To do this, he/she must ask what is necessary, based on the methods used, to produce and distribute the plant, and what are its end uses? van Kester (1990) showed that the market chain for bedding plants is divided into four areas: producers of young plants (plugs and transplants), growers (finishers), distributors and consumers.

For the producers of young plants, breeders must develop a high percentage of good quality plants. Young plants must be uniform and sufficiently vigorous to withstand automated transplanting.

The grower, on the other hand, requires a series of cultivars, which must be able to thrive under a range of growing conditions. It is likely that specific cultivars will be developed in the future for specific climatic conditions, resulting in a wider selection for a more segmented market. For the grower, a uniform response to growing conditions and treatments, within and between a series, is essential. Breeders must also produce cultivars in which flowering time is under the control of the grower and provide cultivars which respond well to low energy inputs.

Distributors of plant materials, however, require cultivars, at both plug and finish stage, to exhibit ease of handling (e.g. lack of brittleness) and the ability to tolerate shipping conditions. Shelf-life, at the distributor and the retail level, is becoming more and more important as the number of retailers proliferates.

The consumer indirectly determines cultivar popularity based on garden performance and appearance in the sales container. However, because consumers do not purchase by specific cultivar, their influence on cultivar popularity is minimal.

The responsibility for bedding plant breeding rests in a small number of hands. The private sector (commercial seed firms) has concentrated on developing more 'grower-friendly' cultivars with reduced cropping time, compact growth habit and enhanced seed quality. The public sector includes universities, research stations and botanical gardens, where plant breeders concentrate on new methods of plant breeding (cytological, genetic engineering) but also work on species with potential but where little commercial breeding has taken place. For example, work at Pennsylvania State University initiated the commercial development of the seed-propagated geranium. Funding for public institutions in the 1990s has diminished and will probably be further reduced; therefore, more responsibility than ever lies with the private sector for creative solutions to breeding problems and for increased funding to public institutions. Numerous advances in bedding plant cultivars have occurred in the last four decades and can only continue with cooperation between public and private sectors of floriculture.

COMMERCIAL METHODS OF BEDDING PLANT PRODUCTION – TRADITIONAL AND PLUGS

Production techniques of bedding plants have come a long way from the days of direct sowing all plants in containers of field soil. While a few fast-growing species are still direct-sown (sunflower, zinnia), two basic production methods have developed in the last 30 years. They have become known as the Traditional Method and the Plug Method. In the traditional method, plants are germinated in open containers and transplanted early to the final container (flat or pot). The bulk of production time is spent in the final container. The advent of plug technology in the 1970s and 1980s forever changed the face of greenhouse production. Plug methodology resulted in significant reduction of space, time and labour to the producer but required highly technical equipment and extremely close attention to detail. From a physiological point of view, significantly less root disturbance occurs in transplanting plug-raised seedlings to the final container compared to the traditional method. The advent of plugs also allowed the grower to fine tune production to provide the optimal environment based on the growth stage of the plant. In the United States alone, it is estimated that at least 75% of the bedding plants sold are grown from plugs and represent about 1.5 billion plants (Dill, 1993).

Although **systems** of production have changed, the basic **concepts** of production have not. Regardless of whether plants are produced in open containers or in plugs, close attention must still be paid to light intensity, photoperiod, temperature, carbon dioxide, irrigation and fertility practices which influence the growth and flowering of bedding plant species.

The Traditional Method

In the traditional method of producing bedding plants, seeds are sown in open trays or occasionally direct-sown in the final container. If sown in open trays, seedlings are transplanted to the final container as soon as they can be handled. Transplanting is usually delayed until at least two true leaves have expanded in order that transplanters handle the plants by the foliage, not the stems. However, plant stress due to handling by transplanters and damage to roots is a serious problem. Plants are essentially uprooted from the germination container and shock to the root systems is not uncommon. Root disturbance can result in a decreased growth rate and flowering time and increased time in the greenhouse. Reducing transplant shock is one of the main reasons why plug systems have been so readily embraced. The time to transplant varies with germination climate, species and cultivar. For example, geraniums may be transplanted after 10–14 days while begonias require 4–5 weeks. All phases of production, from propagation to anthesis, occurs in the same range, often overseen by the same grower. Little equipment, automation or capital investment is required.

The Plug Method

A 'plug' is the term used to designate a single cell transplant. The plug is one of many seedlings in a standard plug tray. The tray, consisting of 50 to 800 cavities, 2 to 5 cm deep, is approximately 28×54 cm (11×22 inches). Germination occurs in the plug tray and plugs remain in the same tray until they reach 'transplantable stage'. This generally occurs when adjacent leaves touch in the tray and when seedlings may be pulled with a 'plug' of soil attached to the root system (3–8 weeks). The plug allows for rapid transplanting with a minimum of root disturbance. McKee (1981a,b) reviewed physiological factors involved in transplanting of vegetable seedlings

Table 1.7. The four stages of bedding plant plug growth (after Koranski and Karlovich, 1989).

Stage[a]	Description
1	Radicle emerges from seed coat
2	Stem and seed leaves emerge
3	True leaves grow and develop
4	Seedlings ready for shipping, transplanting or holding

[a]Stages 1 and 2 encompass germination, stages 3 and 4 make up the growing-on phase.

by traditional means. Methods of hardening off (or preconditioning seedlings), such as drought hardening, growth regulators and antitranspirants, were useful in reducing transplant shock but he found that root disturbance was the single most important factor in causing transplant shock. Transplant shock, in turn, decreased yield and uniformity. Plug production mimimizes this transplant shock.

Seedlings go through various stages of maturity during their time in the plug and the time spent in the plug tray has been categorized to reflect physiological changes in seedling growth (Koranski and Karlovich, 1989; Table 1.7).

The concept of 'growth stages' has been recognized for many years but only recently has the technology (computer software, environmental sensors, automation) been available to growers to provide the optimum environments for each stage, regardless of the size of the plug container. There is no absolute beginning and end to each stage; however, many experienced producers have become adept at recognizing various stages. The recognition of stages allows the producer to be more scientifically in tune with the progress of the plant and subsequent requirements for its optimum growth and flower development.

Advantages of the plug method

One of the main advantages of the plug method is that seedlings may be transplanted much more rapidly than in the traditional method. Also, less experienced people may be used for plug transplanting compared to the traditional method. Many specialist plug producers have arisen, whose sole function is to produce plugs for growers who finish the product. Propagation of plugs is often accomplished in a germination (growing) room or specialized chamber before moving them to the production area. An initial disadvantage in plug production is the investment in new equipment such as seed sowing machines, rolling benches and environmental control for germination and

Table 1.8. An outline of production phases for traditional and plug methods.

Traditional method	Plug method
Germination. Occurs in open seed flat. In general, a single environment is maintained for the entire germination phase. Seedlings remain in open flat until at least two true leaves appear, after which time they are transplanted to the final container. A modification made by some growers is to direct sow in the final container. *Time.* Approximately 10–21 days; however, highly dependent on germination climate, species and cultivar.	*Germination.* Occurs in a many-celled plug tray. The germination phase is separated into stage 1 and 2 (Table 1.7) and the transition between the two stages is closely monitored. The optimum environment may be different for stage 1 and stage 2. *Time.* Approximately 3–5 days for stage 1, 7–14 days for stage 2, depending on species and cultivar (see Table 2.12).
Growing-on. The phase of young seedling growth from transplanting until flower buds are barely visible. Growing-on occurs in the final container.	*Growing-on.* The phase of young seedling growth until they are ready to be transplanted to the final container. The growing-on phase is divided into stage 3 and 4 (Table 1.7). The optimal environment may be different for stages 3 and 4, depending on species (see Table 3.8). Flower buds are seldom visible at the end of this phase. Growing-on occurs in the plug tray.
Finishing. The time from visible bud to sale (usually anthesis). The greenhouse environment is modified to prepare plants for the retail environment. Finishing occurs in the final container.	*Finishing.* The time from transplanting to the final container until sale (usually anthesis). The greenhouse environment is modified to prepare plants for the retail environment. Finishing occurs in the final container.

production. An outline of some of the differences between production phases in the traditional and plug methods is given in Table 1.8.

The greatest differences occur during germination and initial stages of growing-on. Finishing techniques are similar, regardless of production method chosen.

To grow or to purchase plugs

Growers who choose to include plugs in their production cycle must make a decision whether to buy plugs or to grow them. Many growers choose to purchase plugs for finishing because they do not want the capital investment in greenhouse controls and equipment of plug producers. In countries where many specialist plug producers exist and numerous cultivars from plugs are available, purchasing plugs has become very popular. Species such as *Begonia*, *Eustoma*, *Pelargonium*, *Primula* and *Viola*, which have long germination stages or expensive seeds, are particularly popular. Growers who purchase plugs can supplement other production (such as vegetative geranium propagation), reduce problems of germination and allow delay of greenhouse opening compared with growing seedlings. Purchasing plugs results in rapid turnover of plants and the ability to double or even triple traditional volume. The decision to purchase or grow plugs must not only be based on availability of plug sources, but also on the cost of production versus buying. Costs differ between growers regardless of county, state, province or country; however, Table 1.9 presents some production costs (in $US) for a 'typical' grower in the midwestern United States (Carlson, 1992).

Growers who produce plugs do so to reduce cost per unit plant, control quality and produce plugs of cultivars which are difficult to find from specialist plug growers. Plug producers may sell them to other growers or use them themselves.

Table 1.9. The cost per plug tray (US$) of major bedding plant species in various size plug trays – assuming an overhead cost of 3 cents per ft^2 per day (c. 32 cents per m^2 per day). (Adapted from Carlson, 1992.)

Tray size (no. of plugs)			98	200	288	406	512	800
Soil, tray costs ($)			0.65	0.72	0.70	0.55	0.53	0.51
Crop	Seed cost	Weeks						
Geranium	0.067	6	9.32	16.22	22.10	29.86	36.94	56.21
Petunia	0.008	4	2.84	3.72	4.41	5.20	6.03	8.31
Begonia	0.005	8	3.95	4.53	4.95	5.39	5.90	7.32
Impatiens	0.014	5	3.78	5.27	6.49	7.99	9.45	13.46
Pansy	0.029	6	5.60	8.62	11.16	14.43	17.48	25.81
Marigold	0.005	5	2.89	3.47	3.89	4.33	4.84	6.26
Vinca	0.003	6	3.01	3.34	3.55	3.71	3.97	4.69
Coleus	0.005	5	2.89	3.47	3.89	4.33	4.84	6.26
Salvia	0.019	5	4.27	6.27	7.93	10.02	12.01	17.46

International use of plugs

Plug production of bedding plants has become big business since its inception 20 years ago. In 1991, it was estimated that 3 US billion (10^9) bedding plants were produced from plugs in the United States (Lieberth, 1991). Other countries have embraced plug technology with differing intensities. Growers in the United Kingdom, The Netherlands, France and Germany have seen a rapid shift towards plug production, while the future of bedding plant production in Australia appears poised for plugs. The future of plug production in Japan, however, is not as optimistic due to problems of harsh weather patterns and transportation logistics (Lieberth, 1991). Undoubtedly, the future of plug production worldwide will mirror the market strength of bedding plants in general. Where bedding plants are produced in large volumes, the economics of plug production will result in numerous plug growers.

2

PHYSIOLOGY AND PROPAGATION

Physiological studies of bedding plants parallel those of other dicotyledons. Research on bedding plant species has not received the attention of agronomic and bulbous crops. However, attempts to enhance quality and yield and reduce cropping time have resulted in a close marriage of physiological research with production practices. Since most bedding plants are produced in the greenhouse, bedding plant research has concentrated on environmental influences on quality and flowering. Attempts to determine the influence of photoperiod, supplemental light, temperature, fertility, irrigation and growth regulators on growth and flowering of bedding plant species have been undertaken. Studies of these basic environmental inputs continue and greatly influence production methods in the greenhouse. The advent of plug technology resulted in a proliferation of research to enhance seed quality and germination and on the attendant containers, soil, irrigation and environmental inputs to enhance germination and growth in such a system. Plug technology dominates commercial production and therefore a good deal of investigation continues in that area. Research on ways to enhance postproduction life of plugs and finished plants has also been incorporated in greenhouse and retail procedures.

With the exception of vegetative geraniums (*Pelargonium* × *hortorum*) and New Guinea impatiens, bedding plants are mainly propagated from seed. One of the most important recent advances offered by breeders of bedding plant cultivars has been the improvement in seed quality. Numerous firms offer seed which is selected for vigour, pretreated or 'primed' for rapid and uniform germination.

SEED PROPAGATION

The Seed

The seed rests inside a fruit, which is removed after maturation. Bedding

19

plant seeds consist of an embryo resting inside a protective seed coat (testa). A rudimentary root (radicle), shoot (plumule) and leaves (cotyledons) are part of the embryo plant as well as the endosperm, consisting mainly of starch, oil and protein. The seed coat includes the scar (hilum) where the seed was attached to the parent plant, and a small break in the testa (micropyle) which is pervious to water. The testa is often hard and impervious to water and initial entry of water during germination is through the micropyle.

The seed is formed through pollination, but commercial bedding plant seeds often result from controlled hybridization (F_1 hybrids). Accumulation of complex storage products (fats, oils and proteins) is necessary to ensure survival of the germinating seedling. Compounds such as asparagine, glutamine, minerals and sucrose are translocated into the developing seed and reach the embryo by diffusion and active transport. Synthesis of these compounds into storage protein occurs at the direction of messenger RNAs during the last half of the fruit growing period. The accumulation and synthesis of reserve compounds must proceed without interference and reach a minimum level for high quality seeds. Heavier seeds often result in better germination and more vigorous seedlings. Plants used for seed production must receive proper nutrition, moisture, temperature and light as well as be protected from insect and disease pressure, or accumulation of storage reserves will be interrupted. Lack of reserves results in light, poorly developed seeds. This in turn causes high storage losses, poor germination and weak seedlings.

In all seeds, some form of dormancy occurs after separation from the plant. In a general sense, dormancy is defined as 'a temporary suspension of visible growth of any plant structure containing a meristem' (Lang *et al.*, 1987). Primary dormancy evolved to prevent germination under unfavourable environmental conditions and, under natural circumstances, regulates time, place and conditions for germination. Over time, commercialization of bedding plant species has selected for reduced primary dormancy of freshly harvested seed, with enough dormancy to prevent immediate germination and tolerate storage, but not enough to inhibit propagation when conditions are optimized for germination. Dormancy has many causes. **Seed coat dormancy** occurs in some members of Cannaceae, Geraniaceae and Verbenaceae. Seeds are generally scarified after harvest (*Pelargonium*) or soaked prior to sowing (*Canna*). **Chemical dormancy** is a result of chemicals in the seed acting as germination inhibitors. This occurs in species with fleshy fruits or capsules such as those found in Brassicaceae, Ranunculaceae and Iridaceae and in many desert plants. Leaching the seeds through water soaks (*Iris, Anemone*, many perennial species) is often done prior to sowing. A third form of dormancy is **morphological dormancy**, in which the embryo is not fully developed at the time of seed dissemination. The process of embryo enlargement is enhanced by a period of warm temperature. Two groups of embryos occur in herbaceous flower crops (Atwater, 1980). Some species have rudimentary embryos with little more than a pro-embryo in a large endosperm (Ranunculaceae, Papaveraceae) and may contain germination-inhibiting chemicals. Germination may be enhanced by temperature treatments of 15°C or below, alternating temperatures or chemical treatments. Other species exhibiting morphological dormancy have underdeveloped embryos up to half the size of the seed cavity (Apiaceae, Primulaceae). A temperature of approximately 20°C and treatment with gibberellic acid favours germination. The most common form of dormancy in herbaceous plants is **physiological dormancy**. This type of dormancy is usually transitory and disappears after 1 to 6 months of dry storage (Taylorson and Hendricks, 1977; McDonald, 1980; Bewley and Black, 1985). In reality, this type of dormancy is more of a problem for seed testing laboratories than for growers. Control of physiological dormancy appears to reside in the semi-permeability of seed coverings which surround the embryo (Evenari and Newman, 1952). Current concepts to explain physiological dormancy are that germination is controlled by the interaction of endogenous growth regulatory substances such as gibberellins, cytokinins and abscisic acid, and specific environmental inputs, such as temperature and light (Hartmann et al., 1990). Tree, shrub and herbaceous perennial species most commonly show physiological dormancy; however, the great majority of commercially available bedding plant species do not require pre-germination treatments such as cold or plant growth regulators. Germination begins when specific environmental inputs are provided. These inputs have been the subject of intense research efforts. Additional research has focused on seed treatments, such as conditioning or priming, to enhance rate of germination and quality of seedlings.

The process of germination is recognized as the emergence of the radicle from the testa. Certain basic events are necessary for germination. First, water is absorbed through the testa, and especially through the micropyle. If the seed is deprived of water once absorption begins, death usually results. Second, enzymatic activity allows movement of food reserves to the developing embryo. Essentially, this phase involves the interaction of endogenous hormones (gibberellins, abscisic acid) and enzymes (e.g. amylase), although additional metabolic reactions occur. Research has supported the concept that inhibitors (most likely abscisic acid) are present in the testa and the cotyledons of the dormant seed (Black, 1980; Bianco et al., 1984; Gianfagna and Rachmiel, 1986). These disappear in the early stages of germination or may be neutralized by cytokinins, other naturally occurring hormones. During optimum conditions (stratification in some perennial species or warmth in annuals), gibberellins are synthesized or converted to an available form, allowing radicle emergence. Epicotyl emergence occurs later and may require higher levels of gibberellins or be under a separate control system (Hartmann et al., 1990).

For these stages to occur, proper environmental conditions for germination must be provided. In general, temperature, oxygen, light and moisture must be controlled for optimal seed germination, regardless of species.

Environmental Conditions for Seed Germination

Temperature

Temperature is the most important environmental factor that regulates the timing of germination. Dry, unimbibed seeds can tolerate extremes of temperature and seeds of some species can even be immersed in boiling water for a short time to kill disease organisms. Percentage germination and germination rate are affected by temperature. The rate of germination is invariably low at cool temperatures but rises as temperature increases. Above an optimal temperature, at which germination rate is most rapid, a decline in germination rate occurs. As temperature continues to rise, injury and death to the seed may result. Germination percentage, however, remains fairly constant over the mid-range of temperatures, if sufficient time is allowed for germination. Three temperature points (T_{min}, T_{max} and T_{opt}) have been designated for seed germination (Edwards, 1932). T_{min} is the lowest temperature for effective germination, T_{max} is the temperature above which damage or death will occur, and T_{opt} falls within a range at which the highest percentage of seeds germinate at the fastest rate. For most non-dormant seeds of bedding plant species, T_{opt} lies between 24 and 30°C (75 and 86°F). Seeds may also be categorized into temperature-requirement groups, usually related to their climatic origin (Pollock and Roos, 1972; Atwater, 1980; Bewley and Black, 1985; Herner, 1986).

Cool-temperature tolerant seeds germinate over a wide range of temperatures, but tolerate temperatures down to around 4.5°C (38°F). Bedding plant examples include *Bellis*, *Lobularia* and *Lobelia*. **Cool-temperature requiring** seeds require low temperatures and fail to germinate at temperatures greater than 25°C (80°F). This group tends to include winter annuals such as *Viola*, *Primula* and *Delphinium*. **Warm-temperature requiring** seeds will not germinate below about 10°C (50°F). Planting in cold soil can injure the embryo and result in distorted seedlings. Most bedding plants species, especially those native to tropical and subtropical areas, are included in this category. **Alternating temperature requiring** seeds germinate best with about a 10°C difference between day and night temperatures. This has proved to be important with dormant, freshly harvested seeds. Alternating temperatures is a standard practice in seed testing laboratories, but alternating temperatures do not appear to be a requirement for seeds of commercially available bedding plant species. Specific commercial temperature recommendations for bedding plant species are found in the Appendix.

Oxygen

Exchange of gases between the germination medium and the embryo is essential for rapid and uniform germination. Oxygen is essential for respiration in the developing embryo. Supply of oxygen is limiting in excessively wet media resulting from poor drainage, deep planting or poor medium characteristics. In the greenhouse, oxygen in the medium can be significantly affected by the container depth and volume as well as irrigation practices (Fonteno, 1988a).

Light

Light has long been recognized as a germination-limiting factor (Crocker, 1930). In a number of genera (*Amaranthus, Browallia, Digitalis, Epilobium, Lythrum, Nicotiana* and others), some species must be exposed to light in order for germination to take place. Germination of species of other genera (*Cyclamen, Lamium, Nigella, Nemophila, Phacelia, Phlox*), on the other hand, is inhibited by light (Evenari, 1965; Wilkins, 1969). Studies with seeds of lettuce, a light-sensitive species, showed that far-red radiation (c. 700–800 nm) was strongly inhibitory to germination and that red light (c. 600–700 nm) enhanced germination. It was also shown that if red light was followed by far-red light, germination was inhibited. However, if the last exposure was red light, germination was again enhanced (Borthwick *et al.*, 1952, 1954). Such experiments were instrumental in the discovery of phytochrome, the plant pigment system involved in the many facets of photomorphogensis, one aspect of which is the germination of light-sensitive species. It is likely that phytochrome is universally involved in light-promoted seeds (Flint and McAlister, 1935, 1937). Interestingly, a few growers have taken advantage of the light response by constructing seed chambers with incandescent lights draped in red film for more rapid germination of seeds of begonia, impatiens and other light-sensitive species. Another very important feature of photomorphogenesis in the flowering of bedding plants is photoperiod, which is discussed in Chapter 3. Less is known about light-inhibited seeds, but it appears that the phytochrome system is also involved. The far-red region appears to predominate over the red region of the spectrum, and red light has little or no effect on germination.

A good deal of research into light–temperature needs of herbaceous seeds was conducted by Cathey (1969). He classified the temperature and light responses of many species into ten light–temperature groups, some of which are shown in Table 2.1.

Not all species within a genus necessarily respond to light–temperature environments similarly. In fact, significant differences occur among cultivars. For example, when Cathey (1964) tested 58 cultivars of petunias, 18 had an absolute requirement for light while the remainder displayed comparable germination in light and dark conditions. As a result of this work, as well as other research and empirical data, it is recommended that seeds of some species (e.g. petunia, impatiens) should not be covered during germination (Holcomb and Mastalerz, 1985). Khademi *et al.* (1992) showed that some

Table 2.1. The influence of temperature and light on the germination of bedding plant species (after Cathey, 1969).

Germination group	Genera	Comments
Wide temperature range, no light necessary	Ageratum Chieranthus Kochia Tagetes	Seeds germinate from 10 to 30°C without significant decline in germination
Cool temperatures, no light necessary	Cosmos Dimorphotheca	Seeds germinate from 10 to 21°C, poor germination >27°C
Warm temperatures, no light necessary	Amaranthus Celosia Torenia Zinnia	Seeds germinate when temperatures >13°C, temperatures >24°C retard germination
Restricted temperature range, no light necessary	Heliotropium Myosotis Portulaca Tropaeolum	Temperatures above or below specific temperature range retard germination
Wide temperature range, light necessary	Begonia Browallia	Seeds germinate well from 13 to 30°C, absolute requirement for light
Wide temperature range, light beneficial	Cuphea Impatiens Petunia Salvia	Seeds germinate well from 13 to 27°C, erratic and slow germination in the dark
Wide temperature range, dark necessary	Calendula Consolida Schizanthus Tithonia	Seeds germinate well from 10 to 24°C, light inhibits germination

species of impatiens may be germinated successfully in the dark; however, seeds of all cultivars demonstrated higher germination when placed in the light (Table 2.2).

The optimum environment depends on the species and developmental stage of germination. That is, in the first few hours of germination, the seed may be more tolerant of light, temperature or oxygen content than latter stages of germination. In some cases, cultivars within a species may be treated differently. In the commercial sense, most dry-stored seeds of bedding plants may be germinated under normal light conditions encountered in the greenhouse, and most germinate readily in dark seed chambers if moisture and heat are not limiting.

Table 2.2. Germination percentages of five cultivars of impatiens germinated in the light or dark (after Khademi *et al.*, 1992).

Cultivar	% germinating in light	% germinating in dark
'Accent Pink'	97	80
'Accent Rose'	95	52
'Accent Salmon'	89	61
'Super Elfin Coral'	98	25
'Super Elfin Lipstick'	98	92

Seed Treatments to Facilitate Germination

The emergence of mechanical seeders (Fig. 2.1) caused a revolution among seed firms. No longer is it sufficient to breed another red petunia, regardless of how handsome or compact. Today, the seed industry's keyword is seed quality; the plug grower demands higher germination percentages and more uniform and complete seedling stands. The quest for seed with consistent, rapid and high germination, and singulation by mechanical seeders has resulted in numerous commercial seed treatments available to the bedding plant grower.

Refined seed is clean seed that is physically separated by size, shape, weight or density. By removing poor quality seed, the use of refined seed enhances germination. It is recognized that seeds within a species germinate differently, and grading for size (petunia), weight or density (impatiens) is routinely practised by seed firms. Fractions of seedlots which have been separated are performance-tested and sold. Individual fractions may undergo additional separation to fine tune performance. Once separated and tested, they are packaged and sold at a premium. Refined seeds are marketed under various trade names (High-G, High Tech, High Energy, High Vi, Optiseed) and are becoming more common in the seed trade. In general, refined seed is of predictably high (>90%) and uniform germination, with high vigour and energy, and has been most embraced by growers who use plug growing techniques. The trend today is towards all seed being refined in some way; therefore differences in quality and price between seed treatments may soon diminish.

Primed seed, also known as enhanced seed, has been treated to enhance germination. The concept of priming is to control hydration of seeds, thus allowing pre-germination metabolic activities to proceed but preventing actual radical emergence (Heydecker and Coolbear, 1977; Bradford, 1986; Koranski, 1988a). Seeds are usually soaked at 15–20°C in aerated solutions of high osmotic potential for 5–21 days. The main compound used for bedding

Fig. 2.1. A typical mechanical seeder used for plug production.

plant priming is 20 to 30% polyethylene glycol (Mechel and Kauffmann, 1973), although various salts (KNO_3, K_2HPO_4) have also been studied (Bradford, 1986). Water enters the seed coat slowly, beginning the process of germination. However, proper priming does not allow sufficient water to enter the seed and the process of germination is interrupted. At this point, they are 'primed' for germination when they placed in a suitable environment. When the seeds are removed from the gel, germination is arrested. They are air dried, packaged, and marked with a 'sow by' date. Primed seeds germinate more uniformly and over a wider environmental range than standard seeds. They have been most useful for producers with relatively poor environmental facilities. The major benefit of primed seed is not that percentage germinations are increased but that rate of germination is enhanced (Table 2.3). Certain crops such as vinca (*Catharanthus*) and pansy (*Viola*) are difficult to germinate if temperatures cannot be controlled effectively. For example, successful germination of *Viola* is difficult under warm, humid conditions while *Catharanthus* germinates poorly in cool temperatures. Similarly, expensive seeds are also being primed to ensure high germination. Primed seeds of such crops have been received enthusiastically. They will probably continue to attract a premium price because they appeal to a small number of growers and are not produced by all seed firms. Primed seeds of begonia, dahlia, dusty miller, impatiens, pansy, petunia, phlox, primrose, verbena and vinca have

Table 2.3. The influence of seed priming on the rate of germination of pansy (*Viola* × *wittrockiana*) 'Universal Blue', impatiens (*Impatiens* × *hybrida*) 'Super Elfin Orange' (after Koranski, 1988a), and vinca (*Catharanthus roseus*) 'Little Bright Eye' (after Styer and Laffe, 1989).

Germination temperature (°C)	Primed	Germination on day:					
		4	6	8	10	12	14
Pansy 'Universal Blue'							
15	+	0	0	40	59	92	91
	−	0	0	0	0	42	54
27	+	0	19	68	80	90	89.5
	−	0	0	15	43	77.5	79
Impatiens 'Super Elfin Orange'							
17	+	0	0	0	98	99[a]	
	−	0	0	0	71	91	
29	+	67	99	99	99	99	
	−	50	97	99	99	99	
Vinca 'Little Bright Eye'							
27	+	0	48	84	84	88	
	−	0	0	58	60	65	

[a]Data obtained on day 13, the last day of the experiment.

been successfully commercialized. A problem for seed firms who offer seed priming is to devise large-scale equipment for commercial procedures. An additional limitation is the reduced shelf-life. Primed seeds can be stored for only limited periods of time before seed viability is significantly reduced.

The data in Table 2.3 show that pansy germination is accelerated, regardless of germination temperature, while similar, but not as dramatic results, also occurred with impatiens and vinca. Recent work with *Petunia* 'Ultra White' showed that priming seeds with polyethylene glycol ($360 \text{ g } 1^{-1}$) for 3 days resulted in a 44% increase in germination rate and uniformity compared with non-primed seeds (Khademi *et al.*, 1992).

Seed coating is becoming more common with many bedding plant seeds. Raw seeds are covered with a very thin, coloured inert coating (Plate 4) which provides several benefits for growers using mechanical seeders. Firstly, the seeds are visible against the medium, allowing the operator to determine if cavities have been missed or if more than one seed is falling per cavity. This is important when seeds must be singulated or with expensive hybrid seed. Secondly, in the case of some oblong or elliptical seeds, the coating tends to make the seeds rounder, allowing easier control by most seeders. Also, different cultivars can be coated with different colours to avoid mix-up at planting time. In general, all seeds which are coated have been refined through weight or specific gravity, or in the case of marigolds, seeds have been de-tailed. There appears to be no detrimental effects of coating (if properly applied) on germination or subsequent growth.

Seed pelleting is used for small seeds such as begonia as well as other seeds which are not easy to use with mechanical seeders. Seeds of *Ageratum, Alyssum, Antirrhinum, Begonia, Campanula, Eustoma, Lobelia, Petunia, Portulaca, Ranunculus, Senecio* and *Torenia* have been successfully offered to bedding plant growers. The pellet consists of inert colourful materials which dissolve uniformly and rapidly. Most pelleted seeds have been previously refined (*Ageratum, Alyssum, Antirrhinum, Begonia, Campanula, Eustoma, Lobelia, Ranunculus*) or primed (*Petunia, Senecio*), but seeds of other species may be pelleted as standard seed. Although germination may be a few days slower than non-pelleted seeds and seeds are costlier, the ease of use with a seeder and the increase in singulation makes up for such disadvantages.

Somatic embryos, produced through tissue culture, may be available in the next few years to the commercial grower. Somatic embryos are embryos produced from vegetative tissue, and do not require sexual reproduction. Somatic embryogenesis has the potential for producing asexual embryos artificially but also the potential to produce 'synthetic seeds', which would exhibit the same genotype as the parent plant. Mass clonal reproduction and genetic improvement of cultivars may result from somatic embryogenetic programmes. To date, no bedding plant cultivars have been developed from tissue culture programmes, but the potential benefits are obvious and interest is high in commercial firms.

Specialty treatments have also been developed to allow for ease of mechanical sowing. These include removal of the tails of marigold seeds (de-tailing) and defuzzing of tomato seeds to produce a smoother seed surface.

Disadvantages of seed treatments for the grower

While specialty seed has enhanced germination for the producer, there are also disadvantages. Pros and cons have been listed by Styer (1989) and are shown in Table 2.4.

Seed Size

The size of the seed of any species was evolved by the demand for adequate food upon germination and on the necessity of seed distribution. Many species reach a compromise between the two while others lean heavily to one side or

Table 2.4. The benefits and disadvantages of using specialty seed for production of bedding plants.

Pros	Cons
High germination percentage	High seed costs
Improved performance over a wide range of conditions	Limited number of cultivars available, and varying supplies offered
Faster germination	Shorter storage life
More accurate placement by seeders	Slower and lower germination with pelleted or de-tailed seed
Sold by seed count, therefore better inventory control	
Faster plug turns	
Better control of schedules	

the other. The dust-like seeds of *Begonia* and orchids and the 2 kg seeds of coconut vividly demonstrate nature's marvellous variation. In general, plants that have dispersed themselves over large areas of the earth are those with numerous small seeds and adequate food storage. In the production of bedding plants, the size of the seed becomes important in assessing techniques and equipment for sowing. Since most seeds are sold by weight, the number of seeds per gram is critical when determining germination needs and when ordering sufficient seed for the final crop (Table 2.5).

Table 2.5. The number of seeds per gram for selected bedding plant species. Based on an average count among several cultivars (from Nau, 1989).

Genus	Species	Common name	Number of seeds per gram
Abelmoschus	*moschatus*	Annual hibiscus	100
Ageratum	*houstonianum*	Floss flower	7000
Amaranthus	*tricolor*	Joseph's coat	1540
Antirrhinum	*majus*	Snapdragon	6300
Aster	*chinensis*	Annual aster	420
Begonia	*semperflorens-cultorum*	Fibrous begonia	70000
	tuberhybrida	Tuberous begonia	35000
Brassica	*oleracea*	Flowering kale	245
Browallia	*speciosa*	Browallia	4375
Calendula	*officinalis*	Pot marigold	105
Capsicum	*annuum*	Ornamental peppers	315
Catharanthus	*roseus*	Vinca	735

Table 2.5 *continued*

Celosia	*argentea* var. *cristata*	Crested celosia	1190
	argentea var. *plumosa*	Plumed celosia	1365
Chrysanthemum	*parthenium*	Feverfew	7000
	segetum	Corn marigold	7000
Cleome	*spinosa*	Spider flower	490
Coleus	× *hybridus*	Coleus	3500
Cosmos	*bipinnatus*	Tall cosmos	70
	sulphureus	Bedding cosmos	70
Cuphea	*platycentra*	Cigar plant	560
Dahlia	× *hybrida*	Dahlia	98
Dianthus	*chinensis*	Annual pinks	875
Dyssodia	*tenuiloba*	Dahlberg daisy	6300
Eustoma	*grandiflora*	Lisianthus	21840
Gazania	*splendens*	Gazania	420
Gerbera	*jamesonii*	Gerbera daisy	245
Gomphrena	*globosa*	Globe amaranth	400
Hypoestes	*sanguinolenta*	Polka dot plant	630
Impatiens	*balsamina*	Balsam	115
	× *hybrida*	Impatiens	1900
Limonium	*sinuata*	Annual statice	297
Lobelia	*erinus*	Lobelia	35000
Lobularia	*maritima*	Sweet alyssum	7000
Matthiola	*incana*	Stock	665
Melampodium	*paludosum*	Medallion flower	192
Mesembryanthemum	*occulatum*	Ice plant	3745
Mimulus	× *hybridus*	Monkey flower	21840
Nemesia	*strumosa*	Nemesia	3150
Nicotiana	*alata*	Flowering tobacco	8750
Nierembergia	*caerulea*	Cup flower	6160
Nolana	*paradoxa*	Nolana	175
Pelargonium	× *hortorum*	Zonal geranium	210
	peltatum	Ivy geranium	210
Petunia	× *hybrida*	Petunia	9000
Phlox	*drummondii*	Annual phlox	490
Portulaca	*grandiflora*	Portulaca	9800
Primula	× *polyantha*	Polyantha primrose	1000
	malacoides	Fairy primrose	980
Salvia	*splendens*	Annual sage	260
	farinacea	Mealy-cup sage	840
Sanvitalia	*procumbens*	Creeping zinnia	980
Tagetes	*erecta*	African marigold	315
	patula	French marigold	315
Thunbergia	*alata*	Black-eyed Susan	38
Torenia	*fournieri*	Wishbone flower	13125
Verbena	× *hybrida*	Annual verbena	350
Viola	× *wittrockiana*	Pansy	700
Zinnia	*elegans*	Zinnia	140

COMMERCIAL PRACTICES FOR SEED GERMINATION

After dry storage, the seeds of most bedding plant species are quiescent, that is, they are physiologically ready to germinate if the proper environmental conditions are provided. Seeds of geraniums (*Pelargonium*), however, have a hard seed coat and are routinely scarified by seed firms to enhance germination. Sweet peas (*Lathyrus odoratus*) and canna lily (*Canna* sp.) are examples of bedding plants with hard seed coats which should be nicked or soaked by the grower prior to sowing. In general, bedding plant seeds are placed under warm, humid conditions to ensure rapid, uniform germination. Most seeds of summer bedding species germinate between 10 and 25°C, but optimum germination occurs within a more narrow range (warm-temperature requiring). Seeds of hardy annuals sold in the autumn for winter or early spring flowering usually require cooler conditions (cool-temperature requiring).

Stages of germination

In the greenhouse, the process of germination may be divided into two stages (stage 1 and 2, Table 1.7). Specific environmental information, based on the physiological differences between the stages, has been determined for some of the more important species. Close attention to the environmental needs at various stages of germination enhances germination rate and percentage.

Medium

The selection of medium and the selection of plug tray density has become even more important with the acceptance of plug technology. There is no such thing as 'the definitive mix'. Plants have an amazing ability to adapt to almost any medium; however, uniformity of the medium is most important, particularly in plug systems. The germination medium must be well-drained, well-aerated and free of large particles under which seeds may be trapped. The medium should be sterile and have low concentrations or be totally without fertilizer. Some characteristics of successful germination media for bedding plant growth are listed in Table 2.6.

In the traditional method (Table 1.8), seeds are placed on a fine medium in seed trays until germination then lightly covered with sand or coarse vermiculite. Seedlings are then transplanted to a final container containing a different medium, usually within 10 days. The seed trays used in the traditional method contain more medium and generally are deeper than plug trays, therefore seedlings dry out less frequently once removed from the mist, and drainage is enhanced.

In the plug method, the medium used for germination remains for initial production, and is generally used for 4–9 weeks. The plug tray is shallow and

Table 2.6. Characteristics of a successful germination medium.

(a) The ability to hold water, particularly important in early stages of germination (stage 1).

(b) Good aeration for the developing root, particularly necessary after seedlings have emerged (stage 2). Mixes manufactured with different components (e.g. calcined clay, pine bark, perlite) have different porosities. A medium with 20 to 25% air porosity is excellent for petunias, whereas begonias require only 10 to 15% air porosity (Koranski, 1987). In a given mix, additional aeration may be provided by decreasing the amount of water applied (i.e. frequency of irrigation is reduced).

(c) A low electrical conductivity reading (EC < 1.0 mmho using a 1 : 2 ratio of soil : water).

(d) A pH of between 5.8 and 6.5; less than 5.7 may result in iron and magnesium toxicity in some species.

the volume of the medium for each cell is relatively small, compared to the traditional method. The tray density influences the characteristics of the medium used in plug culture much more than the open seed tray used in the traditional method.

Components used for germination include sphagnum peat, sand, perlite, vermiculite, peat–vermiculite or mixes of these ingredients.

Some Properties and Components of Greenhouse Media

The medium has always been a major influence in the growth of bedding plants (see Bunt, 1988), but the popularity of plugs has made the medium even more important. In the traditional method of bedding plant production, a fine germinating medium low in nutrients is used until transplanting to the final container. The germinating medium is used for 10 days to 3 weeks, depending on species. With the advent of plug production, the same medium is employed for the entire plug phase, from 4 to 9 weeks. Bedding plants must be grown in a medium with good drainage and aeration regardless of whether they are being produced in flats or plugs. Most growers use a soilless mix which they blend themselves or purchase. Commercially available soilless media include mixtures of sphagnum peat, horticulture vermiculite, perlite, rockwool, or bark, mixed in various ratios. Numerous properties of the medium influence production practices, particularly aeration and cation exchange capacity. Aeration affects the ability to irrigate, and therefore the ability to apply water soluble nutrients. Cation exchange capacity (CEC) measures the ability of a medium to retain nutrients for plant use. Media which prevent loss of cations by leaching have a high CEC, and those from which nutrients are easily removed have a low CEC. Most soilless media have a low cation exchange capacity and fertilizer supplements must be added on a regular basis.

Sphagnum peat moss is the most common peat used in greenhouse media (others are hypnum, reed-sedge and peat humus). Sphagnum peat has a high water holding capacity, holding up to 60% of its volume in water. It contains 0.6–1.4% nitrogen and decomposes slowly. It is the most acid of the peats, usually having pH of around 3.0–4.0. Most of the peat used in the greenhouse trade is harvested in Canada, the United States, the United Kingdom, Ireland, Germany and Scandinavia.

The consistency of peat plays a much larger role in plug trays than in larger containers. However, peat moss, more than other components, can vary in quality and pathogen content from batch to batch. Growers often purchase an entire year's supply of peat moss at one time to ensure it is from the same harvest and the same batch.

Bark (pine, fir, redwood and some hardwoods) is inexpensive relative to sphagnum peat moss, a component which bark can partially replace in media. Bark must be composted to be of use for greenhouse crops. Non-composted barks have a cation exchange capacity (CEC) of 8 meq $100g^{-1}$, which rises to 60 meq $100g^{-1}$ after composting. Composting results in the destruction of inhibitory compounds, degradation of wood and fragmentation of larger particles into smaller ones. Bark is screened to various particle sizes and particle sizes ranging from 3 to 10 mm are preferred for greenhouse media.

Vermiculite is mined in the United States and Africa as a mica-like silicate mineral. It is expanded after mining and the expanded perlite has a bulk density of 110–160 gdm^{-3}. This lightweight property is desirable in greenhouse media. Each particle contains thin plates lying parallel to each other; therefore, water holding capacity is high and aeration and drainage good. Expanded vermiculite has a CEC of 19–22.5 meq $100g^{-1}$ and contains small amounts of potassium, magnesium and calcium. Vermiculite from the United States is slightly alkaline while African sources can be highly alkaline (approaching pH 9.0 in some cases). It is a desirable component because of its high nutrient and water holding capacity, good aeration and low bulk density. However, vermiculite compresses over time due to force of watering or heavy components, such as soil. The compression greatly reduces aeration.

Perlite is manufactured from siliceous volcanic rock which is crushed and heated to 982°C. Water adheres to the surface of perlite, but is not absorbed into the particles. Perlite is sterile, chemically inert, has a low CEC of 0.15 meq $100cm^{-3}$ and a pH around 7.5. Perlite provides good aeration and is lightweight ($95g\,dm^{-3}$), often substituting for sand, which has a bulk density of 1600–1920 gdm^{-3}. Its inert, stable nature, combined with excellent aeration and drainage, also make perlite a popular amendment for cutting propagation medium, as is used for rooting New Guinea impatiens and geraniums.

Polystyrene foam brings aeration and light weight $(24 \, \mathrm{g\,dm}^{-3})$ to the medium, and is another popular substitute for sand. Polystyrene does not absorb water and has a negligible CEC. It is obtainable as particles from 3 to 13 mm in diameter. Due to ecological concerns, polystyrene has been banned from landfills in many areas and its future as an amendment is questionable.

Rockwool is produced by burning $(1600°C)$ a mixture of coke, basalt, limestone, and sometimes slag from iron production. The resultant liquid is lengthened into fibres and compressed into a pad of specific density. The fibre pad is cut into the desired dimensions. Horticulture-grade rockwool is usually used for vegetable and cut flower production where plants may be grown directly in the cubes or slabs. However, it may also be granulated for use as an amendment for greenhouse media. The pH of rockwool is between 7.0 and 8.0 and is not buffered; a nutrient solution of pH 5.5–6.0 should be used when media are amended with rockwool. The granular form of rockwool has high available water and aeration properties but has a negligible CEC. A blend of equal parts rockwool and sphagnum peat moss has been successful in numerous potted crops.

Other ingredients, such as sawdust, animal manures and treated sewage are also incorporated. Root media need not contain more than three major ingredients, the selection of which often depends on availability and cost. The density of the media as well as cation exchange capacity are powerful arguments for use of specific components. Commercial mixes vary significantly and may be custom blended with almost any desired combination of components.

Fertility of Germination Mixes

Most commercial mixes also contain granular fertilizers, such as calcium nitrate, potassium nitrate, superphosphate, iron sulphate, trace elements and a wetting agent. Individuals and firms may formulate fertility charges differently, but Table 2.7 shows a 'typical' fertility amendment list for a peat–vermiculite medium.

Bedding plant media in which components have been properly blended should have nutrients available to the plant within established ranges. A satisfactory soil analysis, based on a saturated paste extract, is shown in Table 2.8.

Mixing the Media

Many growers purchase pre-mixed media, while others maintain an area for storage of components and mixing the components. Prefilled containers, for

Table 2.7. Additions to soilless media (peat—vermiculite mix) per cubic metre.

Nutrient ingredients	Amount (m^{-3})
Ground dolomitic limestone	2–3 kg
Superphosphate OR	1–2 kg
treble superphosphate	0.5 kg
Calcium or potassium nitrate	0.5 kg
Trace elements	[a]
Wetting agent	90 ml

[a]Various brand names are available, the amount used depends on manufacturers' recommendations.

Table 2.8. A satisfactory soil analysis, based on a saturated paste extract (from Koranski and Laffe, 1988).

Ammonium	<20 ppm
Nitrates	40–80 ppm
Phosphorus	5–15 ppm
Potassium	35–80 ppm
Calcium	50–100 ppm
Magnesium	25–50 ppm
Sodium	<50 ppm
Chlorides	<30 ppm
Sulfates	<100 ppm

germination or finishing, are widely available and eliminate the need for storage and mixing facilities. Although it is not necessary to grow bedding plants in soilless medium, they have become popular because good quality, uncontaminated top soil is often unavailable or is too costly. Bedding plants have traditionally been grown in a mixture of one part native loam, one part vermiculite or perlite and one part sphagnum peat moss for many years with excellent results. If good quality soil is available, consistent and inexpensive, there is no reason that it should not be used. There is no one 'best' growing medium for bedding plants.

The Influence of the Container on Media Properties

Traditionally, seeds were germinated in open containers which accommodated sufficient medium so that water relations in the container did not significantly fluctuate during the germination process. Management of open

Table 2.9. The influence of cell density on potential soil volume (after Hamrick, 1989).

Cells per plug tray	Volume per cell (cm^3)
200 square	11.0
200 round	9.0
242 square	9.0
288 deep square	9.0
288 square	6.4
406 deep square	4.25
406 square	3.4
406 round	3.5
800 square	1.4
800 round	3.5

containers was relatively easy and the seed medium seldom dried out or became waterlogged during germination. The trend towards plug production, however, has resulted in significant problems compared with germination in open containers.

Soil volume of the plug tray

A standard plug tray (in the United States) is approximately 54 cm long, 28 cm wide and 2–5 cm tall. It may consist of 50 to 800 single cells, the most popular sizes being between 200 and 600 cells. The difference in soil volume differs dramatically (Table 2.9), which in turn affects drainage, aeration and water holding characteristics.

Aeration and water characteristics of the plug tray

No medium has fixed air and water holding capacities; both are significantly influenced by the container and the way in which the medium is handled. Fonteno (1988a) showed that water container capacity and aeration of the same medium significantly decreased as the plug density increased (Table 2.10).

Innovations in plug trays continue. Mr Gene Greiling, a large grower from the United States, developed trays with open cavities between the plantlets (Fig. 2.2a). The open spaces between the seedlings result in additional aeration around the plants and reduce the incidence of disease. From Europe, star-shaped plugs have been introduced, which are claimed to increase aeration and produce a better rooted seedling (Fig. 2.2b). Different densities and configurations are available.

Table 2.10. The influence of plug density on the container capacity (water holding capacity) and aeration of growing media (after Fonteno, 1988a).

Medium	288 cells per tray		648 cells per tray	
	Container capacity (%)	Air space (%)	Container capacity (%)	Air space (%)
Canadian peat	84	1.0	84	0.7
Vermiculite no. 2[a]	64	8.8	69	4.1
Vermiculite no. 3[a]	57	0.1	57	0.0
Perlite	53	10.6	59	4.2
Peat–vermiculite (1 : 1)	85	2.8	87	0.5
Peat–vermiculite (3 : 1)	87	2.9	89	0.6

[a]Vermiculite no. 2 is coarse; vermiculite no. 3 is fine.

Height of the cell

It is well recognized that, given the same media in containers containing the same volume, drainage is most directly related to height of the container. The gravitational pull on the water percolating through the medium is greater in the deeper container than in the shallow one. If water does not drain, aeration is reduced. Therefore, the height of the plug cell plays a significant role in determining the aeration of a bedding plant medium, thereby affecting germination. Air space of media in standard 2.5–3.0 cm deep plugs (273–288 cells per tray) with a volume of 4.5 cm^3 ranged from 1.2 to 2.7%, depending on medium tested. If the same cells were deepened to 5 cm, aeration ranged from 4.8 to 10.0% (Fonteno, 1988b). As cells become more dense, the depth of the plug decreases. For example, 128-plug trays from a leading manufacturer are 3.0–5.0 cm deep while 800-cell trays are only 1.9 cm deep (Anon., 1991).

The effects of handling containers

The water content and aeration of all media are affected by the way in which containers are filled and subsequently handled. Fonteno (1988a) showed that the heavy packing of media into containers significantly reduces aeration (Table 2.11). The handling practice of stacking trays on top of each other compresses the media which in turn reduces aeration (Fonteno, 1988a).

Fig. 2.2. An assortment of plug trays used in today's bedding plant operations. (a) Trays with open spaces between seedlings developed by Greiling; (b) star-shaped plugs introduced in Europe.

Table 2.11. The influence of media packing on water and aeration of a peat–vermiculite medium (after Fonteno, 1988a).

Water/air properties	Container size	
	10 cm standard pot	Bedding plant cell (48 per standard tray)
Light packing		
Available water (%)	51	58
Unavailable water (%)	21	21
Air space (%)	15	9
Medium packing		
Available water (%)	52	56
Unavailable water (%)	26	26
Air space (%)	9	4
Heavy packing		
Available water (%)	49	52
Unavailable water (%)	30	30
Air space (%)	4	2

Choosing the right plug tray

As research concerning plant density, soil volume, aeration and water holding capacity continues, producers must still choose which tray is best for their operation. While the debate concerning density continues unabated, the consensus among researchers and growers is that square cavities are better than round cavities. This is mainly due to the larger volume, which reduces fluctuation in water and nutrient content. Plants develop a larger root system with less root binding (Anon., 1990). Star-shaped plugs also reduce binding and are increasing in popularity. Essentially, the cell density of the plug is determined by the crop, the schedule, and whether or not plants must be shipped or are to be used locally.

Germination in the Plug Tray

Work by Koranski using environmentally controlled germination chambers has fine tuned the germination environment significantly since the work of Cathey (see Table 2.1). Koranski and his coworkers examined stage 1 and stage 2 of seed germination and evaluated environmental needs at each stage. In general, they found that after radicle emergence (stage 1), development

Table 2.12. Environmental guidelines for germination of plugs (stages 1 and 2) of various species of bedding plants (after Koranski, 1989).

Genus	Stage 1				Stage 2		
	Temp. (°C)	RH (%)	Light needed	Days	Temp. (°C)	RH (%)	Days
Ageratum	25–28	90–100	No	2–3	22–24	85–90	7
Begonia	25–27	95–100	Yes	6–7	22–26	90–95	21
Brassica	18–21	95–100	No	3–4	17–18	85–90	7
Catharanthus	24–27	90–95 for 3 days 75–80	No	4–6	22–26	75–80	14
Celosia	24–26	90–95	No	4–5	22–24	85–90	7
Coleus	22–24	90–95	No	4–5	22–24	85–90	10
Dahlia	26–27	90–95	No	3–4	20–21	85–90	7
Impatiens	24–27	90–95	Yes	3–5	22–24	80–85	10
Lobelia	24–27	95–100	No	4–6	20–22	80–90	7
Lobularia	25–28	90–100	No	2–3	22–24	85–90	7
Pelargonium	21–24	90–95	No	3–5	21–24	80–85	10
Petunia	24–26	90–95	No	3–5	22–26	75–80	14
Primula	17–20	90–95	No	7–10	15–18	85–90	14
Salvia	24–26	95–100	No	5–7	21–22	80–85	7
Tagetes	24–27	90–95	No	2–3	20–21	80–85	5
Verbena	24–27	95–100 for 2 days 75–85	No	4–6	22–24	75–80	14
Viola	17–20	95–100 for 3 days 90–95	No	4–7	17–20	75–80	7
Zinnia	25–27	95–100	No	2–3	21–23	80–85	7

until the appearance of the stem and cotyledon (stage 2) benefited from less heat and lower moisture around the germinating seed. Table 2.12 provides some insight into their work. The relative humidity (RH) is approximate and was controlled by increasing or decreasing frequency of irrigation.

Many crops tolerate a wider range of temperatures than those given in the table; however, seedling vigour or uniformity may be affected. For example, warm temperatures of 25°C resulted in germination of 'Crystal Bowl' pansy (*Viola* sp.) in about 70 hours, whereas germination required 140 and 120 hours at 18°C and 20°C respectively (Vollmer, 1991). The recommended temperatures of 17–20°C may result in a little longer germination time than with higher temperatures; however, uniformity and quality of seedlings are improved.

Seeds of crops such as vinca (*Catharanthus*), verbena (*Verbena*) and pansy (*Viola*) are particularly susceptible to high moisture levels. A step-wise decrease in irrigation frequency in stages 1 and 2 is recommended for optimum germination. Additional recommendations of additions of 50–100 ppm nitrogen once or twice a week, depending on species, was also indicated by Koranski's work.

Environments for Seed Germination

Successful germination requires viable seeds and optimum temperatures combined with high aeration and moisture levels around the seeds. The optimum environment may be provided in the greenhouse or in specialized germination rooms. In the greenhouse, an area separate from the production area should be designated for seed germination, in which sufficient heat and moisture may be applied. Fog systems have been developed for greenhouse cooling and have been adapted to germination areas; they are more efficient for plug germination than conventional mist systems. Greenhouse systems can convert 99% of the water to droplet sizes between 5 and 20 microns (mm) and can substitute for traditional mist systems in the seed house. The objective is to reduce significantly evaporation and transpiration to the point where water is needed infrequently. Traditional misting systems leave a film of water on the root media which greatly reduces oxygen supply to the seed and radicle. Fogging significantly eliminates this film. A time clock or humidistat is employed to control the frequency of fog. A humidistat is preferable, and by holding relative humidity (or vapour pressure deficit) constant, germination is enhanced. The drawback to fog is the cost and maintenance involved compared with traditional misting systems. Fog systems must be properly engineered, otherwise excessive maintenance may be necessary. A filtration system capable of screening out particles as small as 5 mm is necessary to prevent blockage of fog nozzles. Water quality must be assessed to reduce scale formation and chemical treatments used to reduce slime organisms which can also plug nozzles. Without fog, opaque plastic, either suspended above the seed trays ('sweat tent') or draped over them, can be used to provide moisture and maintain high relative humidity. Heat is supplied from electrical cables, steam or hot water.

The use of the greenhouse for germination is common and relatively inexpensive but does not provide the same fine environmental control that germination rooms allow. Germination rooms (or growth rooms in Europe) are sufficiently large that racks of trays may be rolled in and out at any time. Insulation reduces temperature fluctuation and a fine mist or fog system provides the moisture necessary for uniform germination. The heat may be supplied by fluorescent lamps or by forced air heaters with sufficient air mixture for proper distribution of heat. Air conditioners may be used for cooling when seeds of pansies or other cool-requiring seeds are germinated. Seeds generally remain in the chamber until the first sign of leaf or stem is visible (beginning of stage 2). Allowing too much time in the chamber, particularly if lights are not used, results in significant etiolation of the seedling. Seeds of many species remain no longer than 72 hours before being moved to a shaded area of the greenhouse.

OTHER MEANS OF PROPAGATION

The majority of bedding plants are produced from seed, although a few are propagated by cuttings. The only species propagated from cuttings are New Guinea impatiens, double impatiens, zonal geraniums (*Pelargonium* × *hortorum*) and ivy geraniums (*P. peltatum*). One to two node terminal cuttings are used in all cases, treated with 0.1 to 0.2% indole butyric acid (IBA) in talc and usually placed at 22–24°C in a peat/perlite mix (1:2, v/v). The IBA powder should be applied with an aerated 'gun' which blows powder onto the cut surfaces. This technique negates the need to dip the cut ends in the powder and reduces the spread of disease. Liquid formulations may also be sprayed on to the cut ends. Because plantlets are without roots, high humidity, warm temperatures and low light conditions must be provided until roots begin to form. As the plant matures in the propagation area, the frequency of misting (or fog) is reduced. Roots occur in geraniums within 5–10 days, depending on cultivar; in impatiens, 4–8 days are required.

Fruit crops such as strawberries and grapes grown in hanging baskets are propagated vegetatively and perennial species (*Astilbe, Gypsophila, Heuchera, Hosta*, etc.) are commonly propagated by tissue culture. However, no 'established' annual bedding plants are propagated commercially by tissue culture or other vegetative means.

3

GROWING-ON

Growing-on refers to early growth phases of the young bedding plant until flower buds are visible. In the traditional method, it includes the time from transplanting from a seed flat to the final container, until visible or macrobud stage (see Table 1.6). There is no discrete, absolute time which defines the phase of growing-on. The visible bud stage is simply a convenient, visible, physiological event at which to terminate growing-on and begin the finishing phase. Time spent in the growing-on phase is directly related to the environment, growing practices and the cultivar.

In the plug method, growing-on takes place in the plug flat as well as in the final container. The first part of growing-on is the time from true leaf development until the plants are ready for transplanting from the plug to the final container. This is referred to as stages 3 and 4 (see Table 1.5). The time in stages 3 and 4 is affected by environment conditions and cultivar but is also influenced by the choice of plug flat (seedlings grown in an 800-cell tray spend less time than in a 100-cell tray). In general, flower buds should not be visible (although initiation has probably occurred) when plugs finish stage 4. Once transplanted to the final container, growing-on is treated similarly to seedlings which were started in the traditional method.

NUTRITION AND MEDIA

Fertilization of bedding plants can make a significant difference in the turnover rate and quality of both plugs and finished plants. Greenhouse fertilization is designed to foster rapid, luxurious growth under semitropical conditions. Compared to agricultural practices, an overwhelming amount of fertilizer is applied to greenhouse crops. Nitrogen applications of 1800 kg $acre^{-1}$ $year^{-1}$ (1 ha = 2.47 acres) are routinely applied to potted crops such as chrysanthemum (Nelson, 1991). Excessive amounts of nutrients and imbalance of nutrients may cause significant difficulties. By the same token, nutrients may become limiting due to constant leaching and rapid assimilation by the plant. Although plugs are grown for a relatively short time, they respond dramatically to small quantities of applied nutrients. Similarly, if nutrients are ignored or poorly employed, serious deficiencies, toxicities or imbalances may occur. Many factors must be considered when fertilizer practices are discussed, including water quality, media, form of fertilizer applied and stage of plant growth. To further complicate matters, not all species respond to similar fertilization programmes in the same manner. However, a common misconception in the bedding plant industry is the belief that proper plant fertilization is extremely complex. While many factors do interact, fertilization practices need not be looked upon as shrouded in a 'black box'.

Factors Influencing Nutrition

The method for fertilizing young bedding plants depends on the water quality and the medium in which the plants are grown. The amount of nutrients applied depends on the existing nutrient charge of the medium, its cation exchange capacity and the environment. Plants respond to fertilizer practices very early, regardless of plug or traditional system.

Fertilizer Types and Application Methods

Water soluble fertilizers are the most common form used in bedding plant fertilization. Purchased or mixed on the premises, water soluble blends may be complete mixes of major and minor nutrients (15–16–17, 20–20–20, 20–10–20; N, P_2O_5, K_2O), or an incomplete source fertilizer (calcium nitrate, 15.5–0–0 or potassium nitrate, 10–0–44). Generally, a concentrated mix of fertilizer is prepared in a stock tank from which a fertilizer proportioner delivers a desired dilute concentration of nutrients to the plants. Proportioners may be purchased which dilute the concentrated solution from 1:16 to 1:300. For example, if the proportioner was set to a 1:50 dilution ratio, the fertilizer in the stock tank would be mixed 50 times more concentrated than would be applied to the plant.

Water soluble fertilizers have become popular in the growing of bedding plants because the nutritional status of plants may easily be monitored and adjusted. Water soluble fertilizers may be terminated at any time, increased to provide immediate availability of deficient nutrients or leached from the medium if necessary. The use of proportioners and fertilizer piping throughout the greenhouse range has also made application and mixing relatively simple.

Slow release fertilizers are also used and may be incorporated into the medium used for greenhouse crops. They are formulated to release their nutrients slowly over a given period of time (from 3 months to years). Their advantage is ease of application; once applied little other fertilization need occur. Their disadvantage is that, once mixed into the medium, they cannot be removed. If nutrient imbalances occur or if fertilization needs to be reduced (e.g. hardening of plants prior to shipping), little can be done. If the proper formulations are selected, this is not generally a problem. Some formulations provide only major elements, others also provide micronutrients. The most popular forms of slow release fertilizers consist of plastic-coated spheres of dry, water soluble fertilizers. Water vapour passes through the plastic coating dissolving the nutrients inside. As water continues to enter the capsule, cracks develop allowing the fertilizer to slowly escape. The longevity of the process is determined by the composition and thickness of the plastic coating. Osmocote®, Sierra® and Nutricote® are examples of plastic-encapsulated slow release fertilizers. The release period for Osmocote® and Sierra® is based on 21°C root temperature, that for Nutricote® on 25°C. Some fertilizers such as gypsum ($MgSO_4$) and limestone have limited solubility and are known as slowly soluble slow release fertilizers. When applied to the medium, a small amount becomes available. As the fertilizer is depleted through leaching or plant metabolism, more is released to replace it. MagAmp® (7–40–6) is a popular 3-month-release slowly soluble complete fertilizer, which also contains a high level of magnesium.

Another form of slow release fertilizers used for greenhouse crops are the chelated forms of various micronutrients. Chelates consist of large organic structures that tightly hold copper, iron, manganese and zinc. The micronutrient–chelate complex may be absorbed by the roots whereupon the micronutrient will be released. Chelates release the micronutrients slowly to the medium; therefore roots may absorb these nutrients as they become available. The main reason for the popularity of chelates, however, is their ability to provide micronutrients at a relatively high pH, i.e. in alkaline conditions. Without chelation, copper, iron, manganese and zinc will precipitate and become unavailable at elevated pH due to elevated levels of carbonates and hydroxides. Chelated micronutrients are protected from precipitation. They are used as a temporary measure when a micronutrient such as iron becomes deficient due to high pH. Other more permanent measures should be instituted to reduce pH but the use of iron chelate provides the iron immediately. Chelates are also used for crops which grow better under higher pH and are incorporated into 'premium' fertilizer blends because of their higher solubility compared with non-chelated forms of the same micronutrients. They are expensive and should be used judiciously.

Other forms of slow release fertilizers include sulfur-coated fertilizers and urea formaldehyde (36% N). Neither are used to any extent in the production of bedding plants.

Water Quality

Irrigation water is the most common additive to the young bedding plant; therefore, the quality of the water can significantly affect the performance of the plants. Several properties of water, particularly pH and alkalinity, are important in crop growth and should be measured prior to application of fertilizers.

pH of water

The term pH refers to the logarithm of the reciprocal of the hydrogen ion activity. From a practical point of view, pH is a measure of the acidity or alkalinity of the water and directly affects the subsequent pH of the medium. The pH scale ranges from 1 to 14; a pH of 7.0 is considered to be neutral, below 7.0 acidic and above 7.0 basic. Since pH is a logarithmic measure, a pH of 6.0 is ten times more acid than a pH of 7.0. pH is important because the number of hydrogen and hydroxyl ions affects the performance of some fungicides and growth regulators as well as the availability of various elements from the medium. Water with a pH level between 5.5 and 7.0 should not cause problems for most bedding plants (Carlson *et al.*, 1992). Water which is highly basic may be treated with phosphoric or sulfuric acid.

Alkalinity of water

The alkalinity of water measures the amount of calcium and magnesium carbonate available in the water. The degree of alkalinity affects the ease or difficulty of changing water pH; the higher the measure of alkalinity, the more difficult to lower water pH. In young bedding plants, an alkalinity reading should be approximately $60 \, mgl^{-1}$ ($1.2 \, meql^{-1}$) $CaCO_3$ (Koranski, 1989). If alkalinity is high, addition of phosphoric acid or nitric acid to the water stream converts bicarbonates and carbonates to carbon dioxide gas and water. The amount of phosphoric acid may be calculated by determining the number of milliequivalents per litre ($meql^{-1}$) that must be neutralized (Biernbaum, 1992) to obtain a water pH of approximately 6.0. An example is provided in Table 3.1.

One must be aware that the injection of phosphoric acid not only results in a decrease in alkalinity and pH, but in the addition of phosphorus as well. For every 10 ml of 75% phosphoric acid injected into 1000 l of water, 8.5 ppm of P_2O_5 (5.3 ppm P) are added (Nelson, 1991). If nitric acid (67%) is used, multiply by 81.9 to determine millilitres of acid to be injected into 1000 l of water. Every 10 ml of nitric acid adds approximately 2.2 ppm N to 1000 l of water (Nelson, 1991).

Table 3.1. Determination of millilitres of phosphoric acid to add to 1000 l of water to neutralize excessive alkalinity (7.8 ml 1000 l^{-1} = 1 fl oz 1000 gal^{-1}) (after Biernbaum, 1992).

Alkalinity from water test	6 meq l^{-1} (300 ppm CaCO$_3$)
Desired alkalinity	2 meq l^{-1} (100 ppm CaCO$_3$))
Alkalinity to be neutralized	6 − 2 or 4 meq l^{-1} (200 ppm CaCO$_3$)

Solution: Multiply alkalinity to be neutralized by 54.6 = 218.4. Inject 218.4 ml of 75% phosphoric acid for every 1000 l of water.

NB. If alkalinity is reported in mg CaCO$_3$ l^{-1} or ppm CaCO$_3$, divide that number by 50 to arrive at meq l^{-1}, then proceed as above.

Elemental content

The water stream should also be measured to ensure that elements useful for plant growth and development are available. A desirable range of various elements in water is shown in Table 3.2. A calcium:magnesium ratio of 2:1 is optimal (Vetanovetz and Knauss, 1988).

Table 3.2. Desirable ranges for specific elements in irrigation water. pH 5.0–7.0; soluble salts 0.0–1.5 mmho cm^{-1}; alkalinity 1.0–100 mg l^{-1} of CaCO$_3$. (Adapted from Curtice and Templeton, 1987, and Vetanovetz and Knauss, 1988.)

Element	Desirable range (mg l^{-1})
Phosphorus (P)	0.005–5.0
Potassium (K)	0.5–10
Calcium (Ca)	40–120
Magnesium (Mg)	30–50
Iron (Fe)	2–5
Boron (B)	0.2–0.8
Copper (Cu)	0.0–0.2
Sulfates (SO$_4$)	24–240
Nitrate (NO$_3$)	0.0–5.0
Chloride (Cl)	0.0–140
Fluoride (F)	0.0–1.0
Sodium (Na)	0.0–50
Aluminium (Al)	0.0–5.0
Molybdenum (Mo)	0.0–0.02
Zinc (Zn)	1.0–5.0

Fertilization of the Young Seedling

Soluble fertilizers are composed of elemental salts; thus, the addition of fertilizers involves the addition of salts to the medium. Soluble salts are known to damage roots of seedlings if they become concentrated around the roots; therefore, conservative recommendations often advise that no fertilizer be applied during early seedling growth. However, studies have shown that early dilute concentrations of fertilizer, regardless of production method, result in more rapid growth of seedlings without significant damage. The medium must have an initial low nutrient charge and soluble salt content (< 0.75 mmho cm^{-1}, $1:2$ ratio soil:water).

In plug trays, fertilization may be started in stage 2, but at concentrations no higher than 75 ppm N. The use of potassium nitrate is recommended to avoid ammonia-related complications. Recommendations with petunia and begonia plug crops suggest approximately 50 ppm N of potassium nitrate during stages 1 and 2 (Koranski, 1988b). Normal fertility concentrations (100–300 ppm N) can be resumed at the beginning of stage 3.

In the traditional method, the positive influence of early feeding of bedding plant seedlings was also pointed out (Armitage, 1984). In these experiments seeds were sown in open seed trays, and provided with up to

Fig. 3.1. The influence of early fertilization on French marigold (*Tagetes patula*) 'Queen Beatrix'. Plants were photographed approximately 6 weeks after sowing. The figure shows 0 (1B), 150 (2B) and 300 (3) ppm N applied from emergence.

Table 3.3. Effects of early nitrogen application to geraniums, zinnia and marigolds. Sown in November. (Adapted from Armitage, 1984.)

Nitrogen conc. from emergence to 4 weeks (ppm)	Days to flower[*]	Leaf surface area (cm^2)[*]	Flower buds[†]	Injury (%)
Pelargonium × *hortorum* **'Hollywood Red'**				
0	113 b	613 b	52 b	0
150	118 a	875 a	65 a	0
300	120 a	640 b	42 c	30
Zinnia elegans **'Fantastic Light Pink'**				
0	52 a	362 a	–	0
150	46 b	340 a	–	0
300	45 b	225 b	–	20
Tagetes patula **'Queen Beatrix'**				
0	62 a	395 a	–	0
150	57 b	393 a	–	0
300	58 b	188 b	–	25

[*]Numbers for each species followed by the same letter are not significantly different using Duncan's Multiple Range Test (5%).
[†]Flower buds of *Zinnia* and *Tagetes* not measured.

300 ppm N at emergence (Fig. 3.1). Petunia, begonia and geranium seedlings provided with 150 ppm N showed no damage and were larger in a shorter period of time than those which received no nitrogen or 300 ppm N at emergence (Table 3.3).

In the above work, fertilization was started when seedlings emerged (7–14 days after sowing). Geraniums, for example, are heavy feeders and benefit greatly from such a practice. This is useful when more growth is desired, but feeding too high a concentration may not be beneficial when short, compact plants are needed.

Interaction of elements

An excess of any element can result in phytotoxicity due to that element, but also high concentrations of one element can prevent the uptake of another. Table 3.4 shows some common antagonisms between elements and demonstrates the importance of elemental balance. If a deficiency of a nutrient in the right hand column is identified, it must be determined if an abnormally high level of the antagonistic nutrient in the left hand column also exists. If so, it makes more sense to reduce the antagonist rather than increase the deficient

Table 3.4. Relationships between excesses of elements in the root medium (left hand column) and element deficiency (right hand column). (Nelson, 1991.)

Element in excess	Element inhibited
N	K
K	N, Ca, Mg
Ca	Mg, B
Mg	Ca
Na	Ca, K, Mg
Mn	Fe, Mo
Fe	Mn

element. It is recommended that the fertility programme provide an approximate ratio of 1:1:1:0.5 nitrogen, potassium, calcium and magnesium, and that the iron to magnesium ratio be approximately 2:1 (Sawaya, 1991).

Soluble salts in the medium should be measured routinely (every 7–10 days) and if salt concentrations are too high, leaching should be accomplished.

Measuring Soluble Salts and pH

Specialty laboratories exist to measure soil nutrients, provide elemental leaf analyses and offer nutritional recommendations based on the results of soil and foliar tests. Such facilities allow objective decisions on production practices to be made based on scientific principles. However, the two most immediate concerns of soil nutrition, soluble salts and pH, may be easily measured by the grower in any greenhouse, without the need for sophisticated equipment or expertise.

Soluble salts

Soluble salts (SS) are a measure of soil conductivity and are measured with hand-held or simple bench instruments. Fertilizer elements are carried as salts and if the concentration of salts is too high, root damage can occur. Similarly, if concentrations are too low, sufficient nutrients may not be reaching the root zone. The measurement of SS does not indicate which elements may be too high or low; however, it does suggest that a higher concentration or greater frequency of nutrient application may be necessary (low SS reading) or that soil should be leached with plain water (high SS reading). A solubridge measures the electrical conductivity of a soil solution and is commonly available to the grower. The ratio of soil to water is also important in the

Table 3.5. Interpretation of soluble salt readings for bedding plant species. (Adapted from Mastalerz, 1977.)

Soluble salt reading[a] (mmho cm^{-1})		
1:2 soil:water	1:5 soil:water	Interpretation
0.1 to 0.4	0.08 to 0.3	Satisfactory for seed germination, too low for growing-on
0.4 to 1.8	0.3 to 0.8	Satisfactory range for most established plants, may be too high for stage 1 and 2 seedlings
1.8 to 2.3	0.8 to 1.0	Slightly higher than desired for growing-on
2.3 to 3.4	1.0 to 1.5	Plants usually stunted, slow growth
3.4+	1.5+	Severe dwarfing, death results

[a]To change readings to mho cm^{-1} × 10^{-5}, multiply results by 100.

interpretation of the reading. Most commercial laboratories use a saturated paste; however, greenhouse measurements may use 1:2 or, in the case of soilless mixes, 1:5 medium:water ratios.

DILUTION RATIO FOR SOLUBLE SALT MEASUREMENT
With soilless media, it is difficult for growers to use a 1:2 ratio of soil to water. The paste which results is difficult to work with and a good deal of media must be collected to obtain enough sample. If 1:2 is difficult, consider using a 1:5 ratio. It is easier to obtain enough fluid to measure and interpretation is similar to that for 1:2 ratios (Table 3.5).

Hand-held meters for instant soluble salt readings are available at nominal cost from greenhouse supply dealers and usually come with easy to

Table 3.6. Relative sensitivity of bedding plant genera to soil salts (after Hofstra and Wukasch, 1987).

Very sensitive	Sensitive	Intermediate	Tolerant
Begonia	Aster	Ageratum	Calendula
Celosia	Coleus	Antirrhinum	Centaurea
Impatiens	Pelargonium	Delphinium	Cleome
Viola	Salvia	Phlox	Cosmos
Zinnia	Tagetes	Portulaca	Dianthus
		Verbena	Lobelia
			Lobularia
			Petunia

understand instructions and interpretations. It takes no time at all to make this test; a small price to pay for the difference between crop success and failure.

SENSITIVITY TO SOLUBLE SALTS

Species vary in their sensitivity to salts, ranging from very sensitive to tolerant (Table 3.6).

FERTILIZER MONITORING

One of the most common causes of soluble salt problems is the improper application of fertilizer. This is often the result of poorly calibrated fertilizer injectors. Plants may be receiving an excess or a deficient amount of nutrients because of malfunctioning equipment. The concentration of fertilizer can be monitored by measuring the fertilizer solution being applied to the crop. Injectors are notorious for losing calibration. The solubridge provides a simple procedure for monitoring fertilizer concentrations, as well as calibrating the injector. However, first a word of *caution*. The solubridge must be properly calibrated before fertilizer tracking commences. Many instruments have built-in calibration tests and self-calibrate easily. If not, a reference solution may be made by dissolving 0.744 g of pure potassium nitrate in $1\cdot1$ of distilled water. This should read 1.41 mmho cm^{-1} on the solubridge. If not, the manual should be consulted for calibration instructions.

Fertilizers can be monitored in the following way:

1. Stock solutions of various concentrations of the fertilizer or fertilizer mixture to be used (100 ppm, 200 ppm, 300 ppm, 500 ppm, 600 ppm N) should be prepared by dissolving the fertilizers in greenhouse water.
2. The conductivity of each concentration on the solubridge should be measured.
3. The solubridge readings may be plotted against concentration as in Fig. 3.2. The points should make a straight line when joined. If they do not make a straight line, fertilizer calculations or graphing technique should be checked for errors.
4. A sample of the **same** fertilizer solution from the end of the hose after going through the injector should then be collected. Measure its conductivity on the solubridge and check it against the graph that was previously made for that fertilizer. For example, if the plot was made with various concentrations of 20–20–20, that graph only should be consulted to verify the concentrations of 20–20–20 being applied to the plants. Separate graphs should be made for additional fertilizers used in production.
5. Using Fig. 3.2 as an example, if the conductivity of the fertilizer solution collected at the hose is 0.9, then the nitrogen concentration is 300 ppm. If you think you are applying 200 ppm N, you are in deep trouble, and need to check calculations or the calibration of the injector.

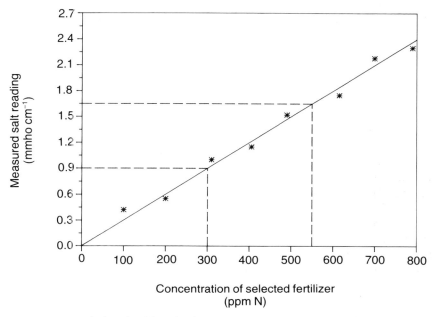

Fig. 3.2. A straight line should result when various concentrations of fertilizer are read on a solubridge (asterisks show actual readings). In this example, if a 300 ppm N solution of the same fertilizer was being delivered, the solution should read 0.9 mmho cm^{-1}; for 550 ppm N, the reading should be 1.65 mmho cm^{-1}. (Adapted from Armitage and Kaczperski, 1992.)

Many chemical supply firms have prepared excellent graphs of conductivity ratings of the various fertilizers they manufacture and can supply additional information on this simple technique.

Greenhouse supply firms also sell injectors with a built-in solubridge. The diluted solution passes through the solubridge and if it is not within the desirable range, amendments to the solution automatically raise or lower the conductivity of the solution. These excellent instruments are commonplace in larger facilities and are becoming more easily obtainable as costs decline.

pH of fertilizers

pH is a measure of the concentration of hydrogen ions in the fertilizer solution and is important because of its influence on nutrient uptake by the crop. When the pH is too low (acid) or too high (alkaline), some nutrients are unavailable to the plant. Research with soilless media (Peterson, 1982) has

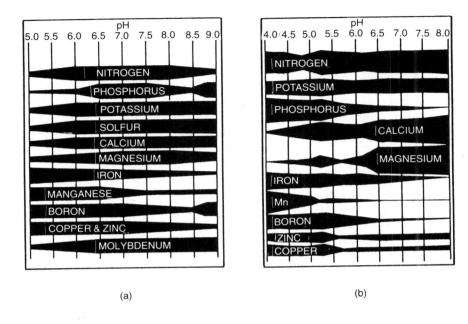

(a)   (b)

Fig. 3.3. The influence of pH level on the availability of essential nutrients in (a) a mineral soil (Truog, 1948) and (b) a soilless medium of sphagnum peat, composted pine bark, vermiculite, perlite and sand (Peterson, 1982).

shown that pH-induced nutrient availability differs between soilless and mineral soils (Fig. 3.3). In general, optimum pH for soilless mixes should be between 5.8 and 6.5 (Peterson, 1984). However, some crops should be grown in soils with a slightly higher pH. For example, geraniums and African marigolds are susceptible to high levels of iron, manganese and zinc. In order to avert a toxicity of these elements, a pH of 6.0–6.5 is recommended. Other crops, such as pansy, petunia, snapdragon and vinca should be grown at a pH of 5.5–5.8 to prevent iron deficiency. Adjustment of the pH is much easier prior to planting than when plants are growing in trays or pots.

Nitrogen Forms

Fertilizer nitrogen is supplied in the form of nitrate-N, ammoniacal-N and urea. Many studies have been conducted to determine the effects of the various forms of nitrogen on plant growth. Some plant nutritionists argue that the nitrate form (NO_3^--N) should be used almost exclusively and that the ammonium form (NH_4^+-N) should be avoided, particularly in the winter. The percentage of NO_3^--N, NH_4^+-N and urea in a given fertilizer is determined by the ingredients used to formulate the mix and can be found on the fertilizer bag. The debate over the use of different forms of nitrogen for greenhouse crops continues but the following facts are known.

Both NO_3^- and NH_4^+ forms can be taken up and metabolized by plants; however, nitrate is often the preferential form for arable crops and is taken up even when ammonium forms are applied. In soils, nitrifying bacteria rapidly convert ammonium to nitrate, which is then assimilated by plants. The response to ammonium and urea is very similar because urea must be converted to ammoniacal-N to be assimilated. Ammoniacal-N releases ammonia which can be toxic to plants, particularly if roots are low in carbohydrates (which occurs under water and temperature stress), but nitrate-N is seldom toxic. The form of nitrogen nutrition can influence the cation (positively charged ions) to anion (negatively charged ions) balance in the plant. Research showed that cations such as Ca, Mg and K were significantly higher while anions such as Cl were lower in plants grown in nitrate-N solution compared to ammoniacal-N (Kirkby, 1968).

The uptake of both nitrogen forms is temperature dependent, rates of uptake being depressed at low temperatures (Clarkson and Warner, 1979). Ammoniacal-N is taken up more readily at low temperatures (10–12°C), but as temperatures rise to 32°C, plants died when supplied solely with ammoniacal-N. No damage occurred between 10 and 17°C (Ganmore-Neumann and Kafkafi, 1983). Temperature, however, significantly reduces the activity of nitrifying bacteria; thus ammonium can build up more rapidly under cool temperatures.

A more rapid uptake of NO_3^- occurs at low pH levels, NH_4^+ uptake is faster at high pH. In general, ammonium fertilizers lower soil pH, nitrate forms cause soil pH to rise. If too much ammoniacal-N is applied, the nitrate uptake can be significantly depressed. In soils, ammonium is converted to nitrate by soil microorganisms; therefore, few instances of ammonium toxicity occur. In soilless media, the level of microorganisms is reduced and the low pH (5–6) in soilless mixes inhibits their action (Nelson, 1991). Therefore, ammonium toxicity can be a serious problem in soilless media. If soil-based media (> 20% soil) are common, few problems with ammonia toxicity are incurred.

No one nitrogen form is best for bedding plants, and arguments that one form or the other should be eliminated are doomed to failure. When all arguments are heard, it may be recommended that best growth and flowering of most bedding plant crops occurs when NH_4^+-N plus urea is less than 50% of the total N applied. This is especially true with soilless mixes during the winter when soil temperatures may be depressed.

Table 3.7. Elemental content of healthy bedding plants. These are guidelines only and not necessarily industry standards. (Adapted from Metcoff, 1992.)

Crop	N (%)	P (%)	K (%)	Ca (%)	Mg (%)	Fe (ppm)	Mn (ppm)	Zn (ppm)	Cu (ppm)	B (ppm)
Begonia	4.4–5.2	0.3–0.4	3.4–4.2	1.3–4.2	0.6–1.0	100–260	90–355	50–65	10–15	30–40
Geranium (seed)	3.7–4.8	0.3–0.6	2.0–4.8	1.1–4.5	0.4–1.0	120–340	110–285	35–60	5–15	35–60
Impatiens (common)	4.3–5.3	0.6–0.8	2.8–1.8	2.9–3.3	0.6–0.8	405–885	205–490	65–70	10–15	45–95
Impatiens (New Guinea)	3.3–4.9	0.3–0.8	1.9–2.7	1.9–2.7	0.3–0.8	160–890	140–245	40–85	5–10	50–60
Petunia	2.8–5.8	0.5–1.2	3.5–5.5	0.6–4.8	0.3–1.4	40–700	90–185	30–90	5–15	20–50
Primula (acaulis)	2.5–3.3	0.4–0.8	2.1–4.2	0.6–1.0	0.2–0.4	75–155	50–80	40–45	5–10	30–35
Snapdragon	4.0–5.3	0.2–0.6	2.2–4.1	0.5–1.4	0.5–1.0	70–135	60–185	30–55	5–15	15–40
Vinca	4.9–5.4	0.4–0.5	2.9–3.6	1.4–1.6	0.4–0.5	95–150	165–300	40–45	5–10	25–40

Tissue Analysis

The best way to determine the actual elemental content of the plant is to send plant tissue to a competent laboratory for testing. Not until the elemental concentrations are known can tissue nutrients be adjusted. Few standards for bedding plants are well known; however, some guidelines have recently been provided (Table 3.7).

TEMPERATURE

In general, warmer temperatures cause faster growth (if measured by dry weight gain). This is true up to optimum temperatures but growth is retarded when temperatures rise above the optimal level. Temperature regulates chemical metabolism and greatly influences such important characteristics as flowering time, flower number, height and dry weight production in plants. Temperature optima differ for different species; however, for most summer bedding plant species, temperatures below 10°C result in slow growth and stunted plants and above 30°C reduce flowering and quality.

Temperature for Bedding Plant Plugs

For plug production, environmental guidelines for some crops for growing on (stages 3 and 4) were determined by Koranski and coworkers (Table 3.8). In general, temperatures are lower than those during germination (stages 1 and 2, see Table 2.12).

Influence of Temperature on Growth and Flowering

Most research on temperature effects on bedding plants was conducted prior to the advent of plug technology. Temperature treatments were often applied 10–15 days after sowing and were studied for various durations, often until anthesis. No one temperature or range of temperatures can be recommended for all bedding plant species; however, air temperatures of 10–25°C are common in greenhouses. As more is understood about stages of bedding plant development, temperature optima may be set differently at each stage.

In general, germination benefits from the warmest temperatures. Temperatures are slightly lowered for seedling growth (stages 2 and 3 of the plug method), and lowered again for plant growth (including stage 4 for plugs) until visible flower bud stage. Finally, temperatures are lowered prior to shipping and sales. Cool temperatures (c. 10°C) result in slow growth, compact plants and delay flowering. This has been shown with geranium

Table 3.8. Environmental guidelines for growing on plugs of various species of bedding plants (after Koranski and Karlovich, 1989).

Crop	Stage 3			Stage 4	
	Temp. (°C)	Fertilizer (N in ppm) and treatment frequency	Days	Temp. (°C)	Days
Ageratum	18−20	150 1 per week	14	15−17	14
Begonia semperflorens-cultorum	21−24	150 2 per week	14	17−20	14
Brassica (ornamental)	17−18	100 1 per week	7	15−17	7−10
Catharanthus	20−22	100 1 per week	21	18−20	7
Celosia	18−21	100−150 2 per week	7	15−17	7−14
Coleus	20−22	150 1 per week	14	15−17	7
Dahlia	18−21	100 1−2 per week	7	15−17	7
Impatiens	20−22	100−150 1 per week	14	15−17	7−14
Lobelia	18−20	100 1 per week	14	15−17	7−10
Lobularia	18−20	100−150 1 per week	21	15−17	7−10
Pelargonium	18−21	150 2 per week	14	15−17	14−21
Petunia	17−20	150 1−2 per week	14	17−18	7−10
Primula	15−17	100 1 per week	35	15−17	7−14
Salvia	17−18	100 1 per week	21	15−17	7
Tagetes patula	17−18	100−150 1 per week	14	15−17	7−14
Verbena	20−22	100 1 per week	14	18−20	7
Viola	15−17	100 1 per week	14	13−15	14−21
Zinnia	18	100 1 per week	7	15−17	7

The fertility recommendations were based on a complete commercial formulation such as 20−10−20 with calcium or a combination of ammonium, calcium and potassium nitrate. Nitrogen concentrations are higher in stage 3 than stage 2. If plugs are to be shipped or stored, reduced nitrogen levels, in the form of calcium and potassium nitrate, are used in stage 4.

(Armitage *et al.*, 1981), petunia (Wolnick and Mastalerz, 1969; Kaczperski *et al.*, 1991) and numerous other species (Seeley, 1985).

Temperature influence on vegetative growth

Both day and night temperatures influence growth of bedding crops. The difference between day and night temperatures influences internode elongation, and controlling this difference has become a useful tool in height control for bedding plant growers. In general, differences in plant height are mainly attributable to differences in internode length, not to differences in node number (Kaczperski *et al.*, 1991). Cool temperatures (*c.* 10°C) result in compact plants with short internodes, while warm temperatures (>25°C) result in plants with longer internodes. Day temperature appears to be more important in influencing plant height than night temperature, regardless

Table 3.9. The influence of day and night temperature on height (cm) of *Petunia* × *hybrida* 'Snow Cloud'. Photoperiod = 18 hours; irradiance = 13 mol day^{-1} m^{-2}; duration of treatments = 10 days from sowing until anthesis. (After Kaczperski *et al.*, 1991.)

Night temperature (°C)	Day temperature (°C)		
	10	20	30
10	11.3	17.5	21.6
20	11.2	25.4	23.0
30	13.3	19.4	25.3

Table 3.10. The influence of day and night temperatures of the length of lateral shoots (cm) of *Petunia* × *hybrida* 'Snow Cloud'. Photoperiod = 18 hours; irradiance = 13 mol day^{-1} m^{-2}; duration of treatments = 10 days from sowing until anthesis. (After Kaczperski *et al.*, 1991.)

Night temperature (°C)	Day temperature (°C)		
	10	20	30
10	8.1	6.0	1.3
20	8.7	4.1	0.8
30	9.7	6.5	0.9

of irradiance. Work by Kaczperski *et al.* (1991) showed that when night temperatures were increased from 10 to 30°C, height increased by an average of 2.5 cm, depending on the day temperature under which plants were grown. However, when day temperature was increased from 10 to 30°C, plant height increased by 11.4 cm under similar night temperatures (Table 3.9).

Lateral shoots are also influenced by temperature. In general, as average temperature increases, the number of lateral shoots decreases (Kaczperski *et al.*, 1991; Kessler and Armitage, 1991). The length (or rate of elongation) of the lateral shoots is highly influenced by day temperature (Table 3.10).

Temperature influence on flowering

Average air temperatures influence flowering time in bedding plants more than the difference between day and night temperature. Recent work with petunias showed that flowering time is a quadratic function of average temperature (Kaczperski *et al.*, 1991; Fig. 3.4).

Work with *Catharanthus roseus* also showed that average daily temperature controlled flowering time (Pietsch and Carlson, 1993; Pinchbeck and McAvoy, 1993) and that the rate of leaf unfolding was also linear

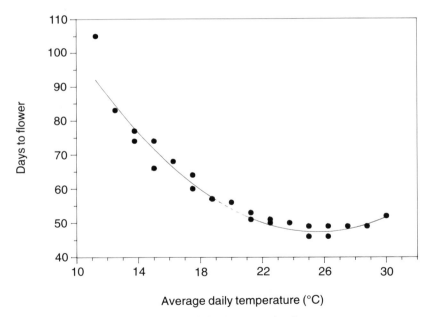

Fig. 3.4. Days to flower for *Petunia* × *hybrida* 'Snow Cloud' in response to average temperature. Photoperiod = 18 hours, irradiance = 13 mol day^{-1} m^{-2}, duration of treatments = 10 days from sowing until anthesis (after Kaczperski *et al.*, 1991).

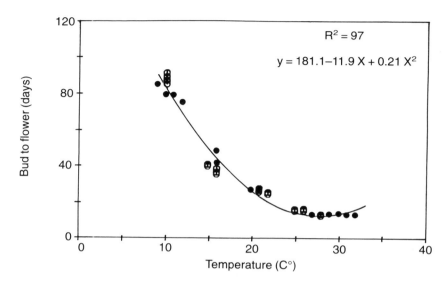

Fig. 3.5. The influence of night temperature on the time between visible bud and anthesis on *Pelargonium × hortorum* (after Armitage *et al.*, 1981).

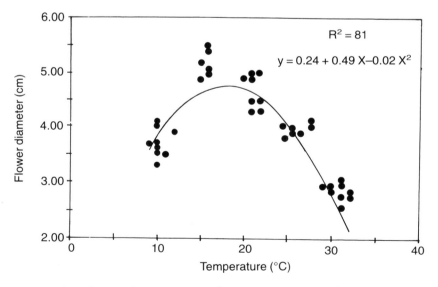

Fig. 3.6. The influence of temperature on flower size in geranium (after Armitage *et al.*, 1981).

between 15 and 35°C (Pietsch and Carlson, 1993). Average daily temperature was a better indicator of flowering time than either day temperature or night temperature alone.

Temperature and irradiance are closely related. Light intensity plays a highly significant role in flower initiation and increased light intensity also results in more reduction in time until flowering at low average temperatures than higher average temperatures (Kaczperski *et al.*, 1991).

Temperature plays a major role in flower development and anthesis. For example, an increase in night temperature decreases the time between visible flower bud (> 0.5 cm) and anthesis in geranium (Fig. 3.5). Since flower organogenesis and development are largely under metabolic control, the importance of temperature at this stage is not surprising and is basic to the rate of flower opening for all bedding plants. Thus, when plants reach the visible bud stage, flowering falls more under the influence of temperature than any other environmental input.

Warm temperatures may also decrease the number and size of flowers in bedding plants, as demonstrated with geranium. As average temperature increased from 10 to 32°C, flower diameter gradually increased, then decreased after 15–20°C (Fig. 3.6). Flower number was not affected until the temperature reached 32°C, at which time floral abortion occurred in the inflorescence (Armitage *et al.*, 1981).

Graphical tracking

The term graphical tracking was coined by Heins and Carlson (1990) as a concept to use temperature to control the final plant height and flowering. Plant height is continuously monitored and temperature, or difference between day and night temperature (DIF, see page 78), is constantly evaluated to maintain the growth and height at acceptable levels. Using a computer to construct the standard graph is not difficult, but is an involved process if done manually. This work is in its infancy but allows continuous monitoring and fine tuning of plant development. Most of the work has been accomplished with potted plants such as lilies, poinsettias and chrysanthemums (Heins and Carlson, 1990).

LIGHT

Three major components of light interact with growth and flowering of bedding plants. They are light intensity, duration (photoperiod) and light quality (colour). Each is important in its own way, affecting flowering, growth and plant quality.

Light Intensity

Light intensity, also referred to as irradiance, photon flux density and photosynthetic photon flux density, is a measure of the amount of useful light striking the plants at any given moment. In general, plants respond to light intensity through photosynthesis with a resulting increase in dry weight. In many crops, particularly non-photoperiodic species, additional light accelerates flowering, resulting in a more rapid crop turnover.

The reduction of sunlight during winter months delays growth of plants. As the season progresses, light intensity and photoperiod increase, resulting in greater cumulative light for the young plant. Typically, winter light in northern European countries and northern states in America may be 20% of summer light (Table 3.11).

The amount of light intercepted by plants in the greenhouse is also related to the construction and the covering of the greenhouse. The frame, sash bars and glass of many greenhouses block out approximately 22% of the incoming light while overhead heating and cooling systems and greenhouse orientation also block sunlight. In many greenhouses, 65% light transmission is not uncommon.

The covering material transmits light at different efficiencies. Clean glass transmits approximately 88% of the light striking it, clean fibreglass reinforced plastic, 88%; two layers of 4 or 6 mil, UV-stabilized polyethylene, 76%; and 8 or 16 mm acrylic panels, 83% (American Society of Agricultural Engineers, 1991). Although double layer polyethylene transmits less light than glass, a double poly house actually may be brighter. This reflects the fewer sash bars and supporting members needed for the lightweight polyethylene compared to a glass house. It is also estimated that 20% reduction in light transmission may occur if coverings, particularly glass, are not cleaned on a regular basis.

Table 3.11. Incoming sunlight (solar radiation) throughout the year as a percentage of July radiation in Madison, Wisconsin (43.2°N, 89.2°W) (after Ball, 1987).

Month	Percentage	Month	Percentage
July	100	January	31
August	87	February	46
September	67	March	61
October	43	April	69
November	27	May	90
December	21	June	92

Flowering time in the north versus the south

In southern areas of the northern hemisphere, seasonal light reduction is similar, but not as dramatic as in northern areas. For example, light intensity in December in Athens, Georgia (34°N, 85.5°W), is approximately 40% of that of July. Although irradiance is reduced in both north and south, the total light received in Athens is greater. Seasonal growth response is obvious in both northern and southern areas in many species. Figure 3.7 demonstrates the response of two bedding plant species as sowing is delayed. The longer that sowing is delayed, the more the young plant is subjected to brighter, longer and warmer days, all of which benefit subsequent growth and flowering.

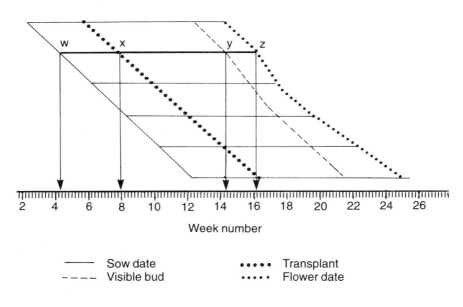

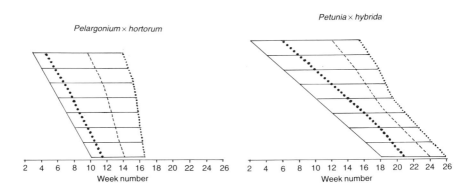

Supplemental growth (supplementary) light

Many people are confused by the term supplemental light. Any additional greenhouse lighting supplements natural light; however, supplemental light can provide two very different functions, depending on the light sources used. The two types of supplemental lighting are supplemental growth light and photoperiodic lighting.

Supplemental growth light is also simply called supplementary light in many countries. It is designed to increase the intensity of light and cause more rapid growth and flowering of bedding plants. Supplemental growth lighting is used because of reduced winter light and is most popular in northern areas of the northern hemisphere. Supplemental growth lights are designed to increase natural light levels and should not be looked upon as the total light input. The sources of supplemental growth light include high pressure sodium, mercury, metal halide and fluorescent lamps, although a number of companies have been working with various hybrids and types. Because the objective of supplemental growth lighting is to increase growth, the spectra of the lamps have been designed to provide light in the photosynthetically active radiation range (400–700 nm) of the light spectrum.

The spectral distributions of some lamps used for greenhouse lighting are shown in Fig. 3.8. The realistic choice of lamps is based on efficiency of energy

---

Fig. 3.7. (See opposite) The influence of sowing date on flowering time of bedding plants. Greenhouse schedules of plants grown by the traditional method giving appropriate sowing and transplanting dates to achieve desired week of flowering (after Armitage, 1983).

Top figure: instructions for interpretation of greenhouse schedules. The axis labelled Week Number refers to the weeks of the year, e.g. point 2 indicates the second week of the year.

1. Choose desired week of flowering (flowering date), shown by a 'z' in the example. Example = week 16.
2. Follow the guideline to the left to the intersection of sowing date 'w'. Example = early in week 4.
3. Follow the guideline to the right to the intersection of transplant week (x). Example = late in week 7.
4. Continue following the guideline to the right to the intersection of visible bud (y). Example = early in week 14.

The above example tell the grower who wishes to market this crop in week 16 that he should sow seed in week 4, transplant in week 7 and see the flower buds in week 14.

Lower figures show schedules for *Pelargonium* × *hortorum* and *Petunia* × *hybrida*.

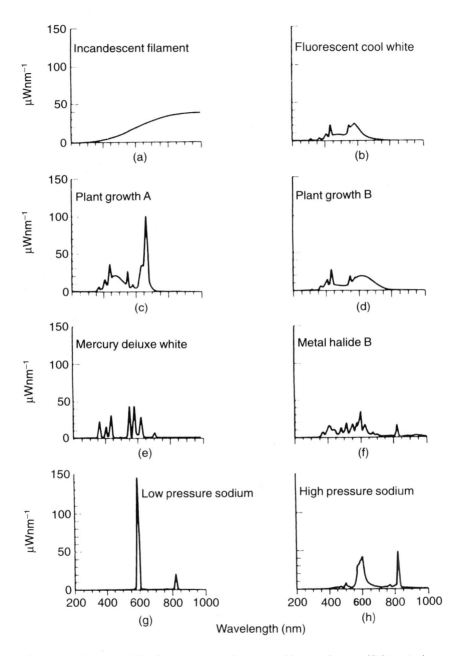

Fig. 3.8. Light spectral distribution of some lamps used for supplemental lighting in the greenhouse (from Campbell *et al.,* 1975).

conversion (i.e. how much light is obtained for the electricity consumed), the distribution pattern (a function of the reflector), cost of purchase and cost of maintenance (lamp life, ballast life, etc.).

Without doubt, bedding plants respond positively to properly designed supplemental growth lighting. Numerous studies have shown that application of supplemental growth lighting under winter greenhouse conditions results in accelerated flowering, more branching, heavier plants and plants of 'better quality'. For example, high intensity discharge lights significantly accelerated flowering of hybrid geraniums (Armitage and Kaczperski, 1992) and begonias (Kessler and Armitage, 1991). Supplemental growth light accelerates flowering time by reducing the vegetative phase of the plant (the time to flower initiation) (Fig. 3.9). The reproductive phase (the time from initiation to anthesis) is less affected by light intensity.

The young plant is most responsive to the application of environmental inputs such as light, temperature and carbon dioxide (Armitage and Tsujita, 1979; Graper and Healy, 1987; Armitage, 1988). If supplemental light is applied, application is most efficient during the first half of the plant cycle. Bedding plant species are often classified as 8, 10, 12 or 14-week crops, referring to the time from sowing to anthesis of first flower. To supplement 10-week crops (impatiens, petunias), growth lights should be used until plants are approximately 5 weeks old. Light may be applied immediately after

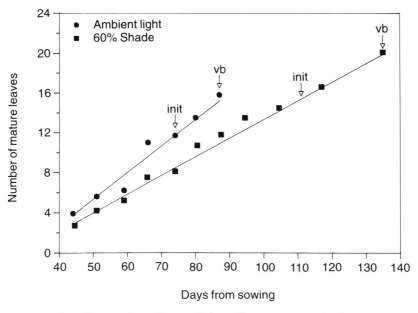

Fig. 3.9. The influence of supplemental light on flower initiation of *Pelargonium* × *hortorum* (after Armitage and Wetzstein, 1984). vb = visible bud.

Table 3.12. The effects of supplemental growth lighting on the percentage increase in dry weight of begonia plugs. Lighting (300–350 foot candles, 18 hours day^{-1}) was started at the beginning of the second week after emergence of the seedlings (Armitage, 1985).

Duration in weeks (and actual weeks after emergence) for lighting	% increase in dry weight compared to unlit plugs
2 (2–4)	5
4 (2–6)	10
6 (2–8)	15
8 (2–10)	16
10 (2–12)	16

transplanting (traditional method) or when plants are in the plug stage. Supplemental light results in more rapid production of dry weight, therefore turnover of the plugs is accelerated. The amount of growth increase due to supplemental lighting is finite, and no additional growth occurs after a certain duration of lighting has been used. For example, lighting begonias for 4 or 6 weeks resulted in almost as much growth increase as lighting for longer periods of time (Table 3.12).

When plugs of *Ageratum*, *Begonia*, *Pelargonium*, *Petunia* and *Salvia* were lit with high pressure sodium lights (300–350 foot candles, 18 hours per day) for 4–5 weeks, increases in dry weight ranged from 10 to 15% compared with unlit plugs (Armitage, 1985).

Photoperiod

The length of the dark and light period is important for flowering and growth in many greenhouse crops. Length of day and night vary with season and latitude, and may have a significant effect on sensitive crops. The change in photoperiod over the year in the northern hemisphere is shown in Fig. 3.10.

While photoperiod is studied and adhered to religiously by growers of crops such as poinsettia and chrysanthemum, little control of photoperiod is practised with bedding plants. That is, most bedding plants grown for spring sale are generally grown under natural day lengths. Producers in the tropics may want to manipulate photoperiod more than those in temperate areas. That is not to say that bedding plant species are unaffected by photoperiod. Cathey (1964) studied the influence of photoperiod on many annuals and his results are summarized in Table 3.13.

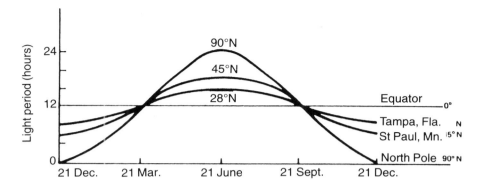

Fig. 3.10. Daylength at various latitudes in the northern hemisphere throughout the year (after Nelson, 1991).

The influence of cultivar on photoperiod

Table 3.13 illustrates that the flowering response of some bedding plant crops may be treated as short day, long day or day neutral. No grower of bedding plants grows the true species today, that is, all bedding plants are cultivars or hybrids. Due to intense breeding over many years, cultivars number in the thousands. Many older cultivars have also been used in subsequent breeding programmes to produce additional new cultivars. In certain crops, this intensive breeding has created cultivars which respond differently to day-length. For example, Carlson (1978) analysed the flowering response of 43 cultivars of *Salvia splendens* to daylength and found that 18 were short day, 14 were long day, and 11 were day neutral plants. In the case of African marigolds (*Tagetes erecta*), many cultivars flower more rapidly under long day conditions (16 hour daylength), while others flower quicker under short days of 8 hour (Tsukamoto *et al.*, 1968; Carlson, 1976). From the production and marketing perspective, the understanding of photoperiodic response allows the grower to force cultivars into flower at different times of year through the use of supplemental photoperiodic lighting or artificial shading. Turnover is more rapid and markets may be extended through judicious use of photo-period to force crops.

Photoperiod may also significantly affect plant habit and must be taken into consideration with certain crops. In many cases, a change in photoperiod results in additional vegetative growth such as basal branching. In the case of petunia (*Petunia* × *hybrida*), plants grown under short days exhibit significantly more basal branching than those grown under long days. The additional breaks result in more flowering stems at anthesis. Thus, petunias which are started early in the year are usually more branched, shorter, but slower to flower than those started later in the season. Growers may take

Table 3.13. The influence of photoperiod on flowering of selected annual species. Some species appear in more than one column due to cultivar differences. (After Cathey, 1964; Carlson *et al.*, 1992.)

Flowering accelerated by[a]		
Short days	Long days	Not affected
Celosia argentea Cockscomb	*Ageratum houstonianum* Ageratum	*Begonia semperflorens-cultorum* Wax begonia
Cleome spinosa Cleome	*Althaea rosea* Hollyhock	*Catharanthus roseus* Madagascar periwinkle, vinca
Coleus blumei Coleus	*Antirrhinum major* Snapdragon	*Gomphrena globosa* Globe amaranth
Cosmos sp. Cosmos	*Centaurea cyanus* Bachelor's buttons	*Impatiens* sp. Balsam, impatiens
Dahlia × *hybrida* Dahlia	*Lobelia erinus* Lobelia	*Lobelia erinus* Annual lobelia
Ipomoea sp. Morning glory	*Nicotiana alata* Flowering tobacco	*Lobularia maritima* Alyssum
Ocimum basilicum Basil	*Pelargonium* × *hortorum* Geranium	*Salvia splendens* Salvia
Salvia splendens Salvia	*Petunia* × *hybrida* Petunia	*Tagetes patula* French marigold
Tagetes erecta African marigold	*Phlox chinensis* Annual phlox	*Viola* × *wittrockiana* Pansy
	Salpiglossis sinuata Painted tongue	
	Salvia splendens Salvia	
	Tagetes erecta African marigold	
	Verbena × *hybrida* Annual verbena	

[a]All photoperiod responses shown are quantitative, that is, flowering is accelerated with the appropriate photoperiod but will occur eventually regardless of photoperiodic treatment.

advantage of the plant's response to photoperiod through proper scheduling to ensure a well-branched, compact plant which flowers in a relatively short time.

The influence of temperature on photoperiod

Temperature can also modify a plant's response to photoperiod. Petunia has been studied extensively and is referred to as a thermo-photoperiodic crop. Under cool temperatures (13°C), petunias are essentially day neutral, between 13 and 22°C they are long day plants, while at temperatures above 22°C they rapidly flower under both photoperiods, although faster with long day conditions (Piringer and Cathey, 1960; Wolnick and Mastalerz, 1969). Other greenhouse crops such as chrysanthemum (*Dendranthema grandiflora*), calceolaria (*Calceolaria herbeohybrida*) and Thanksgiving (or Christmas) cactus (*Schlumbergera truncata*) are also thermo-photoperiodic. Photoperiodic response in many species is not a fixed property and may be altered by changing temperature, light intensity or other inputs.

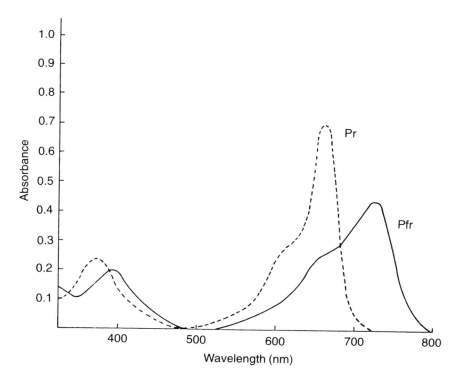

Fig. 3.11. The spectral absorption of phytochrome (after Mumford and Jenner, 1966). P_r = red-absorbing portion of phytochrome; P_{fr} = far-red-absorbing portion of phytochrome.

The control of photoperiod

Control of photoperiod is accomplished through daylength extension, night break lighting or artificial shading. The response to daylength is mediated by the photopigment phytochrome which absorbs light around 660 and 730 nm (Fig. 3.11). Incandescent lights are most often used to provide long days because of their red/far-red spectral characteristics which most closely match the absorption spectrum of phytochrome.

To provide long day conditions, plants are generally subjected to low intensity (c. 100 lx) incandescent light as night break lighting (2–4 hours in the middle of the night), daylength extension (to 18 hours), or cyclic lighting. Short days are generally created by covering the crop with black cloth to maintain darkness for 13–15 hours, depending on species.

Light Quality

Light quality refers to the wavelength of light. Plants respond to the area of sunlight consisting of short wavelengths (<400 nm) of ultraviolet (UV) light to long wavelengths (>700 nm) of far-red and infrared light (Fig. 3.12). The human eye does not see UV light and little of the far-red/infrared spectrum. The plant is most responsive to the visible part of the spectrum, ranging from 400 to 750 nm.

The entire range of visible light is necessary for plant growth and flowering; however the chlorophyll complex absorbs light preferentially in the blue and red regions of the spectrum while phytochrome absorbs light mainly from the red and far-red wavelengths. The plant normally responds to only sunlight, unless receiving additional supplemental light during the winter or photoperiodic light to commence or inhibit flowering. Light spectra for supplemental and photo-periodic light sources have been previously discussed. The role of light quality in height control is discussed in the section on growth regulators.

Because of the possibility of colouring polyethylene, the effects of covering greenhouses with plastics which transmit specific wavelengths have been studied. No beneficial results occurred and, in the commercial production of bedding plants, no spectral quality plastics to influence growth or flowering are used.

SUPPLEMENTAL CARBON DIOXIDE

Carbon dioxide (CO_2) is one of the major inputs of the photosynthetic process. Along with water and sunlight, CO_2 is used by the plant to produce carbohydrates for growth and metabolism. Supplementing CO_2 is no different from supplementing nitrogen or light; it is simply one more input which, if

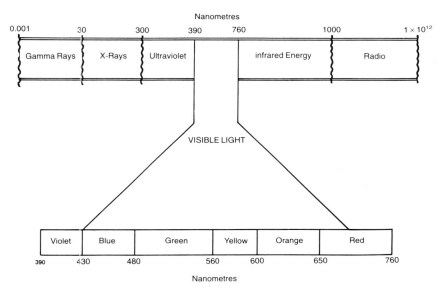

Fig. 3.12. The electromagnetic spectrum.

properly managed, can reduce crop time and improve quality. Numerous experiments on a wide range of crops have demonstrated that, under normal environmental conditions, the concentration of CO_2 limits the photosynthetic rate. Supplementing CO_2 has been shown to be beneficial with china asters (Hughes and Freeman, 1967), chrysanthemums (Hughes and Cockshull, 1971; Hinton-Meade, 1980), carnations (Holley *et al.*, 1964), roses (Goldsberry and Holley, 1962; Lindstrom, 1965) and many other crops.

While additional CO_2 can increase plant growth, there may be times when CO_2 levels in the greenhouse fall below outdoor levels. New greenhouse construction often results in airtight structures, particularly in winter when vents are closed, and little outdoor air enters the greenhouse to replenish CO_2. On bright days, plants are happily photosynthesizing and depleting CO_2 below ambient levels (Fig. 3.13).

While a good deal of greenhouse research has been conducted on CO_2 on vegetables, cut flowers and potted crops, little has been done with bedding plants and few growers practise CO_2 fertilization in the greenhouse. One of the reasons for lack of CO_2 fertilization is that, in warmer areas of the world, the necessity for continuous ventilation limits the effectiveness of CO_2 fertilization. In Canada, northern areas of United States and Europe, however, supplemental CO_2 could be as effective on bedding plants as has been demonstrated on other ornamental crops.

In the greenhouse, supplementing CO_2 to approximately 100 ppm from an hour after sunrise to about an hour before sunset is a useful general recommendation (Hicklenton, 1988). CO_2 systems are controlled by light and CO_2 sensors as well as the opening and closing of the ventilators. As ventilators open to a certain level, CO_2 is turned off. Hicklenton also recommends that CO_2 enrichment cease when ambient light levels fall below 100 $\mu mol\, m^{-2}s^{-1}$ or about 5.4 klx. Light-modulated CO_2 control should be practised whenever possible.

Use of Carbon Dioxide with Plugs

Some of the limitations of using CO_2 on bedding plants grown by the traditional method were that plants were already in their final containers, therefore CO_2 was relatively costly on a per plant basis and that greenhouse treatment was dependent on ambient weather conditions. The economics of supplementing any plant input improve as the number of plants per unit area increases. Nearly all research on supplemental CO_2 on greenhouse orna-mentals was conducted prior to plug technology. While plant benefits could usually be shown, economics of treating pots or flats of plants was not as easily demonstrated. Therefore, the logical next step in CO_2 research was to treat bedding species in the plug stage. Kessler and Armitage (1993) treated begonia plugs with various levels of CO_2 in growth rooms with artificial light. The effects were striking; when provided with 970 ppm CO_2 for 4 weeks, proper light (112 $\mu mol\, m^{-2}\, s^{-1}$), and warm temperature (27°C), the time from sowing to transplant to the final container was reduced by 26% (Fig. 3.14). Two weeks of CO_2 treatment were ineffective.

Three basic questions concerning the use of CO_2 on plugs must be determined when dealing with various species. How much CO_2 is needed, how long must it be applied, and how old must the plants be before they are responsive to the additional CO_2. Recent work with geraniums and pansies (Kaczperski et al., 1993) demonstrated that approximately 1000 ppm CO_2 should be applied to seedlings which are at least 2 weeks old to decrease transplantable time (Fig. 3.14). Applying 1500 ppm was only slightly more effective than 1000 ppm, 4 weeks duration was more effective than 2 weeks but one week was insufficient (Fig. 3.12). Although 2-week old plants only had one or two true leaves, they were as responsive as 4-week old plants with two to four true leaves (Fig. 3.15).

Figure 3.15 shows that if 2 weeks of CO_2 is provided to 14-day-old plugs (7/28), they can be transplanted earlier than control plants. 500 ppm CO_2 was ineffective, but 1500 ppm was of no greater value than 1000 ppm.

According to the data in Fig. 3.16, 14 days of treatment was effective but the most rapid flowering occurred after 3 weeks of treatment. More than 3 weeks did not provide additional benefit.

There is no reason to believe that plugs could not be treated similarly on

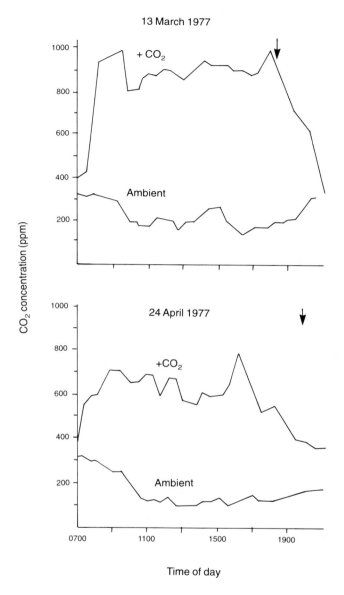

Fig. 3.13. The change in carbon dioxide (CO_2) in the greenhouse over time (from Hicklenton and Joliffe, 1978). Arrows represent times at which vents were opened.

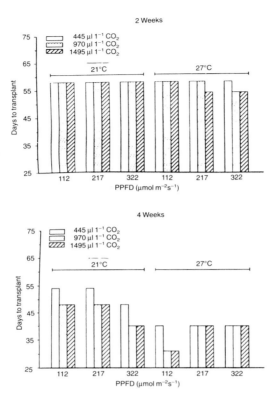

Fig. 3.14. The influence of carbon dioxide on begonias grown in a plug system. The plug stage occurred under artificial conditions (after Kessler and Armitage, 1993). PPFD, photosynthetic photon flux density.

the greenhouse bench, particularly when light is supplemented and vents remain closed. The advent of improved growing systems (greenhouse and growth room) combined with plug technology demand additional research into the use of CO_2 for bedding plants.

The ability to use CO_2 during warm temperature periods has received some attention and met with limited success. Work has included summer CO_2 fertilization in northern countries (Hand, 1984), systems of using CO_2 in southern areas (Kimball and Mitchell, 1979; Willits and Peet, 1981) and pulsing CO_2 to coincide with periods when the ventilators are closed (Enoch, 1984). With the demand for high quality crops and the pressure to turn crops over as rapidly as possible, use of CO_2 will probably increase in the future. Ironically, if the rise in global atmospheric CO_2 due to the burning of fossil fuels does not subside, growers will be using elevated CO_2 without even thinking about it.

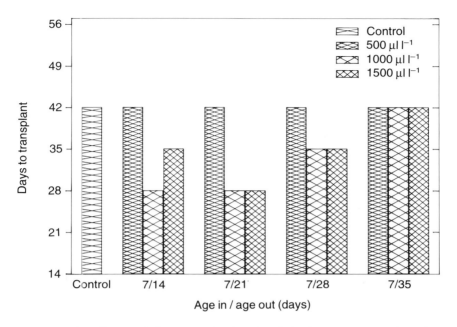

Fig. 3.15. The influence of plant age and carbon dioxide level on time to transplant of *Viola × wittrockiana* 'Super Majestic'. Duration of treatment was 2 weeks, irradiance = 225 μmol s^{-1} m^{-2} (after Kaczperski *et al.*, 1993).

GROWTH REGULATION OF BEDDING PLANT CROPS

In floriculture, environmental and chemical growth regulation are practised to improve rooting response, increase the number of lateral shoots, pinch flower buds, affect flowering time and flower number, and control height. However, in bedding plant crops, growth regulation has principally come to mean height control, although secondary effects on flowering have also been noted.

Height Control

The desire to grow and flower plants quickly has resulted in the use of high levels of fertility, frequent irrigation and warm temperatures, thereby causing internode elongation in many species. Historically, plants grown under cool temperatures and low fertility practices seldom had a problem with height; however, the time required to grow such crops became prohibitive in many

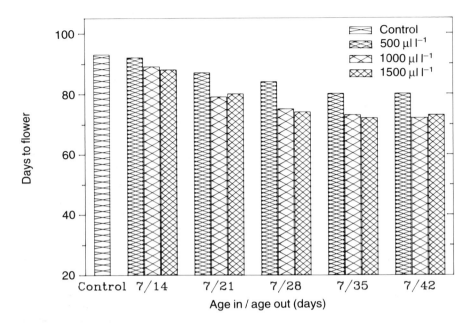

Fig. 3.16. The influence of duration of carbon dioxide treatment on time to flower of *Viola × wittrockiana* 'Super Majestic'. Irradiance = 225 μmol s^{-1} m^{-2}.

areas. Smaller containers also contributed to the perception of tall plants when, in fact, it was the containers that were too small. The wish to squeeze a big plant into a small container, combined with the need to ship as many plants as possible on densely spaced shelves underscores the importance of height regulation. Two common methods for height control used in today's greenhouse are: (i) the use of chemical growth regulators; and (ii) controlling the difference in night and day temperature (DIF). Additional means of non-chemical height control which are potentially important are foliar brushing and manipulation of spectral quality.

Chemical height control

The use of growth regulators has been practised for many years and experimentation with new chemicals continues. Chemicals are applied by spraying the plant or drenching the media. Chemicals are taken into the plant, which in turn restrict internode elongation. Growth regulators block the synthesis of gibberellic acid (Lang, 1970) and the more effective the blockage, the more efficient the chemical and more widespread its commercial use. That activity of gibberellic acid is reduced by growth regulators was shown by experiments when gibberellic acid was applied to plants after application of growth regulators. The effects of the growth regulators were reversed in all cases (El-Zeftawi, 1980; Wample and Culver, 1983; Armitage, 1986a). Growth regulators also affect carotenoids (Greenwald, 1972; El-Zeftawi, 1980) and increase chlorophyll levels (Halfacre *et al.*, 1969; Tezuka *et al.*, 1980; Starman *et al.*, 1989), which tend to make plants greener after application. Significant reductions in water use also occur which can be attributed to reduction in leaf area (Wample and Culver, 1983).

The use of growth regulators is strictly a grower benefit, and in many cases may be detrimental to the plants in the greenhouse or in the landscape. The chemicals used, particularly the new generation, are powerful and easily abused. With most species, the benefit of growth regulators is mainly realized in shipping. The grower who deals with the commercial landscape buyer or who sells directly to the public does not require chemicals and they are not recommended, except in specific cases or for specific cultivars. Although many objections to growth regulator chemicals occur, they are used and are effective, if properly applied (Fig. 3.17).

Many references show the effectiveness of specific chemicals on various species as well as provide excellent overviews (Nickell, 1984; Larson, 1985; Tayama and Carver, 1989). **Depending on the country, state or province, growth regulators may or may not be registered for use. All discussion and recommendations in this book are for information only, and it is the responsibility of the user to check labels for registered use, rates and application frequencies**.

SPECIES RESPONSIVE TO GROWTH REGULATORS

Most growth regulators interfere with the gibberellic acid synthesis pathway in the plant. Therefore, the more efficient the chemical in disrupting this pathway, the more broad-spectrum it is and the lower the concentration needed for effectiveness. The major growth regulators used in today's greenhouses are daminozide (B-Nine), ancymidol (A-Rest), chlormequat (Cycocel), paclobutrazol (Bonzi) and uniconizol (Sumagic). Nearly all species of bedding plants are responsive to one chemical or another and a recent, although incomplete, list of effectiveness of five chemicals is presented in Table 3.14.

Some species are particularly sensitive to certain chemicals. For instance, Barrett and Nell (1991) classify bedding plants' responses to uniconizol (Sumagic) in three categories (Table 3.15).

CONCENTRATION AND TIME OF APPLICATION OF CHEMICALS

Chemicals should be applied at a particular physiological stage, not scheduled by time (e.g. 2 weeks from sowing, etc.). Growth rate is determined by the greenhouse environment, therefore a certain physiological age may occur at

(c)

Fig. 3.17. Photographs showing how application of growth regulators results in shorter plants compared with greenhouse control plants. (a) *Zinnia elegans* 'Peter Pan Red Flame', (B-Nine); (b) *Gerbera jamesonii* 'Jongenelen', (Cyocel, B-Nine, A-rest); (c) *Hypoestes sanguinolenta*, (A-rest, Cycocel).

different times, depending on season. The number of leaves (e.g. first or second true leaf) is often used as a visible marker for initial application, although additional applications are often planned for 7 to 14 days later. The onset of the visible flower bud is also used as a marker for late application of growth regulators.

For spray application, the amount of chemical required (ppm) is calculated and 200 ml of growth regulator solution per square metre of growing area (1 gal per 200 ft²) of growing space should be applied, depending on chemical. Chemicals should not be sprayed to 'runoff'. This refers to spraying until chemical runs off the leaf to the bench or container. Not only is this a waste of chemical, but significant damage to plants may occur. 'Runoff' is particularly dangerous to plants when using Bonzi and Sumagic, where volume of chemical is equally as important as concentration. Davis (1991) recommends 120 ml of Bonzi or Sumagic solution per square metre (2.5 quarts per 200 ft²) due to potential damage from runoff.

For drench application, the diluted solution is prepared and then

Table 3.14. Effectiveness of five growth regulating formulations (trade name, with chemical name bracketed) on various bedding plant genera (after Carlson *et al.*, 1982; Barrett and Nell, 1991).

Genus	B-Nine (daminozide)	A-Rest (ancymidol)	Cycocel (chlormequat)	Bonzi (paclobutrazol)	Sumagic (uniconizol)
Ageratum	+	+	−	+	+
Antirrhinum	+	+	−	+	+
Begonia (fibrous)	−	+	+	+	+
Begonia (tuberous)	−	−	+	−	?
Callistephus	×	×	×	−	?
Catharanthus	+	×	×	+	+
Celosia	+	+	+	−	+
Cleome	+	+	+	−	?
Coleus	−	+	−	+	+
Dahlia	+	×	+	−	?
Dianthus	−	×	+	+	?
Impatiens	+	+	−	+	+
New Guinea impatiens	−	−	−	−	+
Pelargonium	−	+	+	+	+
Petunia	+	−	−	+	+
Salvia	+	+	+	+	+
Tagetes	+	+	+	+	+
Verbena	+	+	+	−	+
Viola	−	−	+	+	+
Zinnia	+	+	−	−	?

+ = effective, − = not effective or response not known, × = response varies by cultivar, ? = response not known (for Sumagic only).

Table 3.15. Response of specific bedding plants to uniconizol (Sumagic) (after Barrett and Nell, 1991).

Sensitive <5 ppm ai	*Begonia, Catharanthus, Pelargonium, Viola*
Moderate 5–20 ppm ai	*Celosia, Coleus, Impatiens,* New Guinea impatiens, *Salvia, Tagetes, Verbena*
Least sensitive >20 ppm ai	*Ageratum, Antirrhinum, Petunia*

ai, Active ingredient.

56–112 ml of solution per 10 cm of pot size (0.5–1 oz of solution per inch of pot size) is applied. Drenches should be applied in the morning on actively growing plants or plugs to avoid phytotoxicity.

APPLICATION TO TRANSPLANTS

In the traditional method of growing bedding plants, growth regulators are generally applied after transplanting to the final container. For example, after three to five true leaves have formed (approximately 5 to 6 weeks after sowing), chlormequat is usually applied to seed geraniums. An additional application may be provided 7 to 10 days later. Similar schedules ensure that application of chemicals occurs prior to stem elongation but after roots are well established in the container. Most recommendations for traditional culture of bedding plants cite only daminozide, ancymidol and chlormequat, since paclobutrazol (Bonzi) and uniconizol (Sumagic) were not available until after plug culture was well established. Table 3.16 shows typical recommendations for the three established growth regulators on plants grown by the traditional method.

APPLICATION TO PLUGS

Since plants may remain in plug trays for up to 10 weeks, depending on species, growth regulators are used during the plug stage. The need for height control for plugs is similar to that for finished plants, that is, to reduce plant stretch and facilitate shipping. Bonzi and Sumagic have been added to the arsenal of chemicals on many species, but their potency and difficulty of application have limited their popularity.

The rate of application of growth regulators and the timing is dependent on the physiology of the seedling as well as the environment in which it is growing. The growing temperature, irrigation and fertility practices are particularly important. When temperatures in stages 3 and 4 are maintained below 13°C, growth is slow, internodes are close together and the need for growth control is diminished. Similarly when fertilizers are applied, stretching inevitably occurs with warm temperatures and high levels of nitrogen. Plugs which have been grown with little irrigation and fertilizer and grown at cool temperatures are referred to as 'hard' plugs. Generally, these plants stretch less than 'soft' plugs (grown warm with optimum levels of water and fertility for rapid growth) and require less growth regulator. 'Hard' plugs are generally more compact but require more time to finish than soft plugs.

Optimum chemical rates are lower for plug-grown than for traditional-grown plants. Barrett (1989) recommends about one-quarter to one-half the amount used on cell packs or 10 cm pots. Repeated low rates result in less damage than one or two concentrated applications. The first treatment is generally made when the stem begins to elongate in late stage 2 or early stage 3. Subsequent applications may be made at 5 to 7-day intervals. Sawaya (1993) provided guidelines for growth regulators on plugs (Table 3.17).

Table 3.16. Growth regulator recommendations for concentration, application method and timing for bedding plants grown by the traditional method (i.e. plugs not used).

Material	Concentration and method [a]	Application time
A-Rest (ancymidol)	33–132 ppm ai, spray	Apply 2–4 weeks after transplanting to final container
B-Nine (daminozide)	2500–5000 ppm ai, spray	Apply 2–4 weeks after transplanting to final container. May be repeated at 2–4 week intervals
Cycocel (chlormequat)	750–1500 ppm ai, spray; 2500–4000 ppm ai, drench Used for *Pelargonium* and *Dianthus* only	Apply when plants have two to four true leaves. Spray may be repeated at 1–2 week intervals, drench only applied once

ai, Active ingredient.
[a]Spray or drench.

Table 3.17. Growth regulators effective on annual plugs. Based on work by Sawaya (1993). Spray volume was 50 ml per cell tray. The temperature was 19–21°C and photoperiod was 10–14 hours.

Species	Concentration (ppm ai)				
	Sumagic	Bonzi	A-Rest	Cycocel	B-Nine
Ageratum	1	4	1		2500
Antirrhinum		4			
Begonia (tuberous)	1			250	
Celosia	1	4			
Coleus	1	4	1		5000
Dahlia	1	5	1		5000
Dianthus	1	4		250	
Impatiens	1	4			
Pelargonium (seed)	1			750	
Petunia	1	4			2500
Portulaca	2	8			5000
Salvia	1	4	1		
Tagetes	3	12			
Viola	1	4	1		
Zinnia		12			2500

ai, Active ingredient.

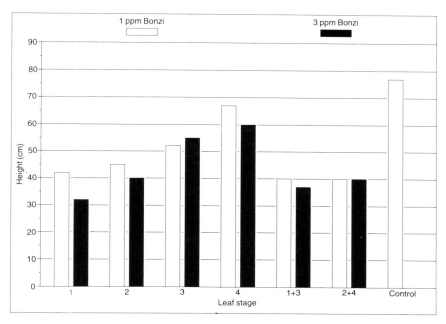

Fig. 3.18. Effect of Bonzi on height of *Viola × wittrockiana* 'Universal Orange' applied at different leaf stages (after Laffe and Styer, 1989). The authors also showed that treatment of plugs at the first leaf stage as well as multiple applications resulted in stunted plants.

Additional work done by Laffe and Styer (1989) with paclobutrazol demonstrates the importance of timing on height control of pansy plugs (Fig. 3.18). Two application rates were evaluated by applying at different leaf stages.

OTHER RESPONSES TO CHEMICAL GROWTH REGULATORS

The basic tenet for application of chemical growth regulators to bedding plants is to restrict plant height. However, side-effects due to growth regulation have been noted with a number of species. Most growth regulators affect chlorophyll synthesis, resulting in greener plants. Improper application of chlormequat (too much or applied when too warm) often results in severe marginal leaf chlorosis (colour plate 5). This type of damage limits the use of chlormequat and has resulted in growers applying frequent, less concentrated applications rather than one or two highly concentrated sprays. Repeat applications at lower concentrations are generally more effective than single concentrated applications and cause less damage, regardless of chemical selected.

Growth regulators have mixed affects on flowering. In many species, flowering is unaffected or slightly delayed (see Fig. 3.17a,b). However, Cycocel, when applied prior to floral initiation in seed-propagated geraniums, results in significant reduction in flowering time (Miranda and Carlson, 1980) (Fig. 3.19).

In most species, flowering is usually delayed if chemicals are applied after flowers have initiated. Fibrous begonias (*Begonia semperflorens-cultorum*), however, produce additional flowers when treated with chlormequat. An undesirable observation with paclobutrazol is the long persistence of the chemical, resulting in delayed growth of plants in the transplant container or the landscape. The growth delays noted by growers and landscapers are probably the result of improper application, not the fault of the chemical itself. No prolonged persistence occurs with the first generation of chemicals (daminozide, ancymidol, chlormequat).

A subtle postproduction benefit may also be observed in the retail environment. This is that plants are more compact, and therefore they lose less water and are less susceptible to stress (see section on postproduction).

As with any chemical, growth regulators must be applied correctly or significant damage may occur. Growth regulators are simply a tool for growers, like the fertilizer bag or the thermostat. Properly used, they enhance plant quality; improperly used, they can be a disaster.

Fig. 3.19. Photograph showing how, regardless of the number of applications of Cycocel applied, flowering of *Pelargonium* × *hortorum* 'Sprinter Scarlet' is accelerated.

Non-chemical height control

Some of the problems associated with chemicals (leaf chlorosis, delay of flowering or growth) have prompted growers to attempt to slow plant growth by non-chemical means. Ironically, even with the powerful chemicals developed for height control, many growers continue to modulate growth through the reduction of temperature, water and fertilizer. This was the main means of maintaining plant architecture prior to the advent of growth regulating compounds, and still continues today. Such practices result in slow growth and flowering and plants remain in production longer than when grown with optimal temperatures, water and fertility. If grown properly in such a restrictive manner, excellent quality plants may be shipped without problem. The knowledge of how much and how often to restrict such important environmental parameters requires experience and an excellent eye for detail.

DIF technology

The effects of differences between day and night temperatures on plant growth have been reported by many workers including Cathey (1954), Went (1957), Herklotz (1964), Cockshull *et al.* (1981) and Parups and Butler (1982). Experiments with day/night temperature manipulation took on additional significance during the world energy crisis in 1973. Although experiments found that energy savings could be realized by reducing night temperature, the differences between day and night temperatures resulted in excessive stem elongation and plant height in a number of important greenhouse crops compared with plants grown at constant temperature or cooler day than night temperature (Tangeras, 1979; Moe, 1983). A relatively large body of research has shown that plant morphogenesis in a wide range of species is influenced by the difference between day and night temperature (DIF). DIF has a significant effect on internode elongation and subsequent plant height. If plants were subjected to day temperatures higher than night temperatures, called a positive DIF (+DIF), internodes tended to stretch, but if subjected to day temperatures lower than night temperatures (negative DIF, – DIF), internodes were short. Moe and Mortensen (1992) summarized much of the European and American research on DIF and made the following observations:

1. Average daily temperature (ADT) has a negligible influence on stem elongation.
2. The difference between day and night temperature strongly influences internode length and total plant height in a wide range of pot plants.
3. The response to DIF is quantitative; the more positive the value of DIF, the greater the response (Erwin *et al.*, 1989, 1991).
4. The magnitude of the response to a change in DIF is not the same across

all values of DIF; internode elongation increases more as DIF increases from zero to a positive value than from a negative value to zero (Moe, 1990).

5. DIF has the greatest influence on plant height during the stage of rapid growth.

Additional work suggests that DIF may be mediated or interact with phytochrome. Work with *Campanula* showed that a low red to far-red ratio (incandescent light) nullified the negative DIF effect whereas a high red to far-red ratio (fluorescent light) enhanced the effect of negative DIF (Moe *et al.*, 1991). Gibberellins are involved in stem elongation and it is likely that DIF influences gibberellin synthesis and/or action, in much the same way as red/far-light responses and chemical growth regulators. The response to DIF is greatest at moderate mean temperatures and is reduced at high or low mean temperatures (Vogelezang, 1992). Many bedding plants respond to the application of DIF, such as *Antirrhinum, Celosia, Dianthus, Hypoestes, Impatiens, Pelargonium, Petunia* and *Salvia*. French marigolds (*Tagetes patula*),

Table 3.18. The influence of DIF and growth regulators on four bedding plant species (Vogelezang *et al.*, 1992).

Day/night temp. (°C) and growth regulators	Plant height (cm)[*]	Flowering time (days)[*]
Pelargonium × ***hortorum* 'Pulsar Red'**		
13/21	16.5 a	82.6
16/18	16.8 a	83.5
18/16	19.2 b	82.5
+ chlormequat	13.9 a	80.2 a
− chlormequat	21.1 b	85.6 b
***Petunia hybrida* 'Blue Flash'**		
13/21	21.5 a	56.9
16/18	25.6 b	56.8
18/16	23.8 ab	53.0
+ daminozide	18.4 a	58.0 b
− daminozide	28.8 b	53.1 a
***Impatiens* 'Impulse'**		
13/21	8.3 a	
16/18	12.2 b	
18/16	14.0 b	
+ daminozide	10.8 a	
− daminozide	12.2 b	

[*]Statistics analysed separately for temperature differences and growth regulators for all taxa. Figures followed by the same letter for the same species.

on the other hand, show little response (Heins and Erwin, 1990). Table 3.18 shows some data on the influence of DIF on four European bedding plant species.

The data in Table 3.8 show that elongation of *Impatiens* and *Petunia* were sufficiently regulated by reversed temperatures if a large negative DIF (−8) was received, however, a −2 DIF was not effective. In *Pelargonium*, −2 DIF was effective but chlormequat was still needed. However, the authors demonstrated that the amount needed was reduced by 30% with reversed temperatures. Flowering time was not affected by DIF treatments.

Controlling height by DIF is particularly attractive when chemicals of all types are continually scrutinized and disappear from agricultural production. Reducing dependence on chemicals not only makes environmental sense but also financial sense. The reduction in growth regulator cost can be significant. Many growers have used DIF recently and significant responses have occurred with many crops. Not all species are responsive, but enough are to suggest that the technology should be attempted wherever practical.

The day to day implementation of DIF to control height can be handled through graphical tracking (Carlson and Heins, 1990). Essentially, the height of the crop is monitored each day against a standard height for the crop at a given set of environmental conditions. If the height is too tall or too short, the DIF may be altered to bring the plants back to the standard height. Computer technology is the best way of handling graphical tracking and standard growth curves for most major crops should soon be available as computer software.

LIMITATIONS OF DIF TECHNOLOGY

Effective temperature range. The range of temperatures within which day and night can be manipulated is defined by the crop's response to temperature. Most crops respond to rises in a temperature range by more rapid growth and earlier flowering. However, outside this range, metabolic processes are adversely affected and extreme temperatures cannot be used. This may be a problem when night temperatures must be raised to achieve a large negative DIF. Figure 3.20 shows a generalized effect of temperature on the relative growth rate of many plants.

Similar temperature sensitivities occur during flower initiation and development. Such sensitivities are greatest at the time flowers are initiating or at early flower development. Once flower buds are visible, DIF has less of an effect on final flower development (Nelson, 1991).

Side effects. As with any newly implemented technology, the widescale execution of DIF has been accompanied by numerous problems. One is the necessity for a good system of greenhouse control. Environmental computers, while not essential, make temperature control much easier. However, management must pay much closer attention to details if the DIF technology

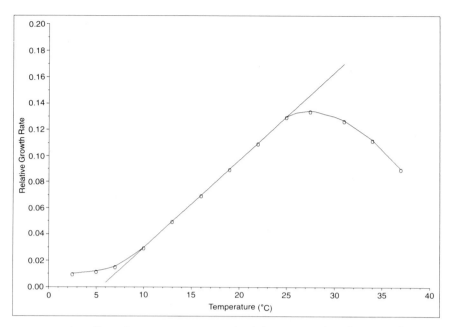

Fig. 3.20. The effect of temperature on growth of plants. Most plants have a similar relationship with temperature. The most effective temperature range is shown by the straight line (after Nelson, 1991).

is to be effective. Lack of attention to temperature or improper manipulation can result in nutritional problems and poor quality plants.

The lower the value of negative DIF (bigger negative number), the greater the chance of chlorosis in immature leaves. If a negative DIF is applied too early, chlorosis can become a permanent problem. This is because the young plug is little more than immature leaves and the chlorosis results in severe plant stunting, which may haunt the plant for many weeks. It is not recommended to treat seedlings with negative DIF values below −2 or −3°C in the first 3 weeks of growth (Nelson, 1991). Chlorosis on older plants generally disappears as the negative DIF is raised. Salvia and gerbera are particularly susceptible to DIF-induced chlorosis.

Downward leaf curl has been a problem with certain crops when low negative DIF has been applied. Raising DIF values makes such problems disappear.

Cool morning/warm evening strategies. In southern countries and states in the northern hemisphere (and also northern areas of the southern hemisphere), and in the summer on radiation-rich days, the implementation of DIF technology is more difficult. To maintain short internodes, a negative

DIF is usually called for. Because day temperatures are almost always elevated in radiation-rich days, it is difficult to maintain a lower temperature during the day than the night. To achieve similar effects of negative DIF on such days, a temperature drop after sunrise has been recommended (Erwin *et al.*, 1989). Venting the greenhouse at dawn for 2–3 hours is an effective means of lowering day temperatures. Erwin worked with Easter lilies (*Lilium longiflorum*) which were highly responsive; however, many other crops do not respond to a morning temperature drop. Hendricks *et al.* (1992) found no response with *Chrysanthemum frutescens*, *Fuschia hybrida* or *Pelargonium zonale*. Vogelezang *et al.* (1992) also found little response to cool morning strategies with *Begonia*, *Fuschia*, *Impatiens*, *Pelargonium*, *Petunia*, and *Salvia*. On the other hand, Moe and Heins (1990) and Moe and Mortensen (1992) found that pot plants such as poinsettia and Eatior begonia responded with reduced internode elongation when provided with a 2 hour cool morning treatment. Obviously, some species specificity occurs, but with the vast diversity of plant species used in floriculture, this should not be surprising. There is little doubt that implementation of DIF technology is most practical in areas of relatively cool day temperatures during the growing season.

Heaters are often turned on for a few hours before sunset and warm air vented to reduce humidity and disease. This warm evening strategy has little effect on stem elongation (Hendricks *et al.*, 1992).

Other non-chemical methods

MANIPULATION OF LIGHT QUALITY

Growth of many species may be altered through manipulation of the red to far-red ratio. Crowding of plants on a bench is thought to result in taller plants with longer internodes because, as light passes through the leaf canopy, the ratio of red to far-red light is decreased (Kasperbauer, 1971). Increasing the amount of red light relative to far-red has been shown to inhibit growth of greenhouse crops (Warrington and Mitchell, 1976; Mortensen and Stromme, 1987; McMahon *et al.*, 1991), peas (Noguchi and Hashimoto, 1990) and maize (Vanderhoef *et al.*, 1979) seedlings. The manipulation of light quality may be an effective means of height control for areas where DIF or chemical control is not practical. One way to manage the red/far-red ratio is through the use of copper sulfate ($CuSO_4$) solution over the crop canopy. Copper sulfate increases the red to far-red ratio as sunlight passes through. Recent work by Rajapakse and Kelly (1992) with chrysanthemums showed that $CuSO_4$ reduced plant height by reducing internode length. Stem elongation rate and total leaf area were also significantly reduced compared to control plants (Table 3.19).

The reduction in growth could be reversed through application of gibberellic acid, suggesting that the reduction of stem elongation is mediated by changes in gibberellic acid synthesis or action. Lockhart (1964) suggested

Table 3.19. The influence of copper sulfate on chrysanthemum growth after 4 weeks of treatment (after Rajapakse and Kelly, 1992).

| CuSO$_4$ conc. | No. of leaves | Reduction compared to control (%) | | | |
		Plant height	Internode length	SER[a]	Leaf area
0	17	—	—	—	—
4	16	34	33	48	25
8	15	42	40	60	39
16	15	37	29	54	32

[a]SER = stem elongation rate (cm week^{-1}).

that conversion of gibberellic acid to the 'active form' may be prevented by red irradiance and prevented by far-red irradiance. Little work has been accomplished with bedding plants; however, Benson and Kelly (1990) found significant reductions in height (40–66%) when pansies, petunias, geraniums and impatiens were grown under copper sulfate solutions. There is nothing to suggest that bedding plant species would not respond in a greenhouse environment.

This work and other earlier work suggests that greenhouse coverings may be manufactured with specific spectral characteristics to influence plant height. Israeli scientists worked with experimental plastics which absorbed light in the 530–570 nm range and emitted it at 630–670 nm (Raviv, 1989). Kadman-Zahavi and Ephrat (1976) found that plants grown under plastics which transmitted blue and far-red light were taller than those grown under plastic which transmitted primarily blue light. In experimental greenhouses, copper sulfate has been pumped through double acrylic panels to reduce the heat load in the greenhouse, allowing vents to remain closed so that carbon dioxide could be introduced even in the summer (Mortensen and Stromme, 1987). Follow-up studies in Norway examined five coloured solutions in fluid roof chambers (Mortensen and Moe, 1992). They found that the copper sulfate solution caused the most dramatic reduction of internode length. This could be explained by the high red to far-red ratio and blue to red ratio compared to the control water solution (3.8 vs. 1.0 and 1.88 vs. 1.00 respectively). A yellow solution removed most of the blue light and caused significant shoot elongation. Control of plant growth may be possible with filters and manipulation of incoming radiation through filters, solutions or plastics of varying colours. Presently, the practice has not yet been adopted on a commercial scale.

FOLIAR BRUSHING

Wind is a natural growth regulator in nature. It reduces stem length and leaf area (Bryer, 1967; Jaffe, 1973), and the idea of brushing to emulate the effect of wind has merit. Mechanically induced stress, such as rubbing, bending, shaking or brushing the stems or shoots of plants results in several growth responses, the most obvious being a reduction in stem and petiole length causing the plant to develop a shorter, more compact growth habit. Most work has been conducted on greenhouse-grown vegetable plants (Biddington, 1985), in particular cauliflower, lettuce, celery (Biddington and Dearman, 1985) and broccoli (Latimer, 1992). Additional research on the influence of foliar brushing (thigmomorphogenesis) on height has been conducted with tomato, eggplant and some cucurbits (Latimer, 1992). Brushing 60 times with a wooden slat over the top 2 to 7 cm of plant growth resulted in height control and reduction in petiole length with little foliar damage (Latimer, 1992). The technology for large-scale greenhouse brushing has not been tested and little other research on ornamental bedding plant species has been reported.

FINISHING

The finishing phase of bedding plant production refers to the time once flower buds are barely visible (c. 1 cm) to the time the plants leave the greenhouse. This may be less than a week to 3–4 weeks, depending on species, cultivar and environment under which the plants are grown.

Scheduling Times

Scheduling of crops is essential to greenhouse profitability. While high quality and reasonable price are essential in selling plants, they also must be ready when the market wants them. The demands of the market, the climate, the season and the cultivar influence the dates of sowing, transplanting and sale. In general, temperatures are monitored closely so that anthesis occurs on schedule (see Fig. 3.7). If the crop is on schedule, then temperature, watering frequency and fertility are often reduced during the final flowering phase (from visible bud to anthesis). This is referred to as hardening off or toning the plants to prepare them for the postproduction stage of their journey. The chapter on postproduction provides additional information on grower responsibilities to lengthen postproduction life.

While there is no one schedule for any crop, a marigold is still a marigold, regardless of whether it is grown in New York or New Zealand. The schedules in Table 3.20 are presented as guidelines only, knowing full well that the inside and outside environment greatly influences their accuracy. They have

Table 3.20. Schedules for bedding plants grown in traditional and plug methods. Schedules are based on a midwestern United States climate and are meant as guidelines only. From Bodger (1985), Koranski and Karlovich (1989) and Nau (1989).

| Crop | Traditional method | | Plug method[a] | | |
	Days from sowing to transplant	Total crop time (weeks)[b]	Weeks in plug	Weeks from transplant of plug to sale	Total crop time (weeks)[b]
Ageratum	15–20	10–13	5–6	4–5	9–11
Begonia semperflorens	45–50	13–17	8–9	5–7	13–16
Brassica	10–15	6–7	3–4	4–6	7–10
Catharanthus	30–35	11–14	6–7	6–8	12–15
Celosia	10–15	9–12	5–6	4–5	9–11
Coleus	20–25	7–9	5–6	4–5	9–11
Dahlia	11–15	11–13	3–4	3–4	6–8
Impatiens	15–20	9–11	5–6	3–4	8–10
Lobelia	20–25	9–12	5–6	5–8	10–14
Lobularia	20–25	7–9	5–6	2–3	7–9
Pelargonium	10–15	13–15	6–7	8–11	14–18
Petunia	15–20	11–13	5–6	2–4	7–10
Primula	40–45	16–20	9–10	10–14	19–24
Salvia	14–18	9–10	5–6	4–5	9–11
Tagetes patula	10–15	8–9	5–6	2–4	7–10
Verbena	20–25	10–13	5–6	5–7	10–13
Viola	15–20	11–15	6–7	6–8	12–15
Zinnia	10–15	7–8	3–4	3–4	6–8

[a]Plug method assumes the use of 406 plugs.
[b]Crops finished in cell packs, 32 or 48 per tray.

been selected from various sources of literature including seed company publications, as well as from presentations given by growers and technical representatives.

4

POSTPRODUCTION

There is a large difference between the benevolent climate in the greenhouse and the sales area or garden. The terms 'hardening off' and 'toning' are often used to describe the acclimatization process for greenhouse crops. Hardening off has received the most attention in the field of vegetable transplants because of their significant change from greenhouse to field conditions. Withholding water, subjecting plants to cold temperatures, modification of the nutritional status, use of chemicals such as antitranspirants and other seedling treatments have been studied to reduce transplant shock. Hardening off is effective in reducing losses along the bedding plant production–retail chain; however, the physiological effects from hardening off are poorly understood. Physiological and metabolic changes have not been studied in bedding plants and most of the work has been conducted with vegetable transplants (Levitt, 1980; McKee, 1981a,b). Physiological changes which occur from hardening include a decrease in transpiration rate (Kramer, 1950; Bieloral and Hopkins, 1975) and an increase in endogenous hormone levels (Wright, 1972; Hsiao, 1973). Loomis (1925) found that hardened plants recovered their transpiration rates more rapidly than non-hardened plants, recovery being correlated with an increased rate of shoot and root growth. The rate of photosynthesis also decreases during hardening, mainly as a result of increased stomatal resistance. However, hardened plants may recover photosynthetic rates after stress more rapidly than non-hardened plants (Boyer and McPherson, 1975). It has also been shown that during stress, average soluble carbohydrates within hardened plants were increased (Eaton and Ergle, 1948; El-Leboudi *et al.*, 1980).

Many of the same treatments are of limited use in hardening off bedding plants in order to better withstand stresses of shipping, retail handling and transplanting to the garden. Hardening off should begin in the greenhouse and continue during shipping and retail handling. Significant decline of quality can occur at the shipping and retail level if plants are poorly handled. The advent of plugs decreased transplant shock by root disturbance but they too benefit from hardening off at various stages of production. This involves the holding or 'storage' of plugs and finished plants.

POSTPRODUCTION OF PLUGS

Hardening off Plugs

Plugs go through a definite postproduction phase. After 4 to 6 weeks of production, many are shipped to another grower for finishing. Preparing plugs (hardening off) for shipping is the responsibility of the grower. Cooling plugs in the greenhouse and reducing irrigation frequency and fertility prior to shipping is practised by many growers. Research on hardening has been conducted with flowering (finished) bedding plants and interest in plug shipping and movement has resulted in research activity in this area as well. The reasons for hardening off plugs are the same as for finished plants; to prepare the plants for a more difficult environment. In the case of plugs, shipping can be stressful; for finished plants, shipping, followed by the retail environment, can be extremely difficult.

Holding Plugs

Holding occurs when plugs or plants are ready for transplanting or sale but transplanting is delayed because of personnel, facility or weather problems. Therefore the plugs or plants must be held until they can be transplanted or sold. The practice of holding plugs should be discouraged, but realistically this is not always possible. Late season sales or overproduction may also result in a backlog of plugs at the plug grower level or at the finisher. Cold temperatures are the key to holding plugs and may be accomplished on the greenhouse bench or in a cold storage facility.

Plug storage in a cool greenhouse

The temperature of temporarily empty greenhouse facilities may be lowered to 8–10°C to successfully store petunias, geraniums and begonias for up to 4 weeks. Warmth-loving plugs such as impatiens can effectively be stored at 13°C in plenty of light (Firth, 1992). Roll-out tables that transport plugs outdoors during the day and back in at night have also been used. The key to storage in cool greenhouses is to maintain cool temperatures as long as possible, provide good air circulation, and water only moderately, especially if plugs are to be stored longer than 1 week.

Plug storage in a cooler

Recent evidence shows that postproduction of plugs may be enhanced by placing transplantable plugs in a cooler at 8°C in the dark. This has been shown to be effective at the grower level with geranium plugs (Firth, 1992). Additional research has demonstrated that cooling plugs (Heins and Lange, 1992; Kaczperski and Armitage, 1992; Heins and Wallace, 1993) and the reduction of fertility prior to shipping (Kaczperski and Armitage, 1993) resulted in better postproduction life of plugs, whether they were held by the finisher or not.

If plugs are stored by the finisher after shipping, all but the most tender crops may be held at 4–8°C. This includes petunias, marigolds, geraniums, salvia and pansies. Some species such as *Lobularia maritima* (alyssum) can be stored below 5°C (Heins and Wallace, 1993). Others such as impatiens (bedding and New Guinea types) should be stored at warmer temperatures (no lower than 8°C). Table 4.1 provides two cooling temperatures and a tentative list of species for each one. All species should be stored at low light levels (*c.*5 μmol m^{-2} s^{-1}), although dark storage is feasible for short lengths of time.

From the commercial application, storage in the dark is preferable and easier. Heins and Lange (1992) studied the storage limitations of many species and found that a compromise temperature of 7.5°C resulted in approximately 3 weeks of dark storage. Although some plugs (petunia, pansy)

Table 4.1. Recommended storage temperatures for various bedding plant genera. Based on Armitage and Kaczperski (1992), Heins and Lange (1992) and Heins and Wallace (1993).

Storage temperature	
4–8°C	8–12°C
Ageratum	*Begonia* (fibrous)
Antirrhinum	*Catharanthus*
Begonia (tuberous)	*Celosia*
Dianthus	*Coleus*
Lobularia	*Impatiens*
Pelargonium	*Impatiens* (New Guinea)
Perennials (all)	*Zinnia*
Petunia	
Salvia	
Tagetes	
Viola	

may be stored in the dark for up to 6 weeks, a low light level is preferable. Addition of 5 µmol m^{-2} s^{-1} of light resulted in significantly better quality of plants after storage (Heins and Lange, 1992; Heins and Wallace, 1993) than darkness. Warmer storage temperatures may also be used if light is provided. For example, geraniums store well at 5°C in the dark, but if 5 µmol m^{-2} s^{-1} of light were added, they could also be stored at 12°C (Heins and Lange, 1992). Special cold-tolerant fluorescent lamps or incandescent lights should be used. Incandescent light, even under cool temperatures, can result in internode elongation. Fungal diseases may also be a problem under storage conditions; therefore it is recommended that all plants be treated with a fungicide prior to placing in the cooler. If possible, plants should be stored under low humidity conditions. Irrigation is necessary in the cooler if plants are to be stored more than 1 week.

POSTPRODUCTION OF THE FINISHED BEDDING PLANT

Figures from the United States have estimated that losses as high as 20% occur after the finished plants leave the greenhouse. Such astronomical losses cannot be tolerated if the bedding plant market is to remain profitable. While it is easy to blame individual links of the postproduction chain, the responsibility of enhancing postproduction life must be shared by everyone, the grower, the shipper/trucker and the retailer.

Responsibility of the Grower

The responsibility of building postproduction life into the crop begins with the grower, prior to the plants leaving the greenhouse. Many specific activities can be tailored to increase shelf-life at the end of the production phase and will be discussed in more detail; however, any practice which decreases plant quality during production also decreases shelf-life. Such things as leaf chlorosis, marginal, foliar burn, presence of insects and/or disease, poor root growth, or physiological disorders all reduce plant quality. Regardless of the causes of these problems, the shelf-life of a plant suffering stress, albeit unintentional, is reduced. If poor quality plants leave the greenhouse, then the best display/retail area in the world will not improve their postproduction life. The grower plays a well-defined and important role in the postproduction life of the final product. One of the most important functions of the grower is hardening off the plants before shipping. This includes manipulation of temperature, fertility and irrigation prior to shipping or sale.

Hardening off finished plants

In the quest for the fastest turnover and flowering time, crops are often provided with warm temperatures, high fertility rates and constant irrigation throughout the crop cycle. Faster flowering cultivars and technological advances in production, such as constant liquid feed systems, boom watering, and sub-flat heating systems, have not only allowed more critical control of energy, water and fertilizer but have also provided the means to 'push' the crops to maturity as rapidly as possible. In most production manuals for bedding plants, there is little emphasis on hardening off any more, yet if plants fall apart at the retail outlet, repeat sales for next year's fast crop may be slow in coming.

Hardening off is not difficult, needs to be done only for a short period and saves money. It must, however, be scheduled into the production cycle and crops need to be arranged in greenhouses so that the proper environment can be set for this purpose.

HARDENING OFF WITH TEMPERATURE

Since most bedding plants are grown at relatively warm temperatures (15–25°C, depending on season and species), low temperatures are useful to slow growth and flowering. In the greenhouse, reducing both day and night temperature is effective, but reduction of night temperature prior to finishing can provide additional shelf-life to many bedding plant species. This is usually accomplished around the time the flower bud becomes visible (5 to 30 days before anthesis, depending on species and environment). When finished marigolds were grown at 10, 15 or 20°C night temperature from visible bud until market, those held at 10°C had the longest shelf-life (Table 4.2).

In the 20°C retail temperature, marigolds persisted 7 days longer when hardened off at 10°C greenhouse temperature, compared with a greenhouse temperature of 20°C. This also dramatically illustrates the effect of the retail

Table 4.2. Effect of lowering production temperature on shelf-life (days marketable) of marigolds. (Adapted from Nelson *et al.*, 1980.)

Retail temp. (°C)	Production night temp. (°C)		
	10	15	20
10	17	14	12
20	17	13	10
30	4	4	4

Fig. 4.1. Impatiens were provided with temperatures of 10, 21 or 32°C (50, 70 or 90°F) from visible bud until maturity. The photograph shows plants 11 days after storage at 20°C.

Table 4.3. Effect of lowering production temperature on the shelf-life (days marketable) of impatiens. (Adapted from Nelson *et al.*, 1980.)

Retail temp. (°C)	Production night temp. (°C)		
	10	15	20
10	12	15	10
20	13	15	15
30	2	2	2

temperature on postproduction life. When the same experiment was conducted with impatiens, however, the coldest hardening temperature was not the best (Fig. 4.1, Table 4.3).

Impatiens is a warmth-loving species compared with marigolds, so such results are not unexpected. Not all finished crops can be hardened off with the same temperature but two groups can be identified (Table 4.4).

For plants in Group A, lowering temperature results in significantly less

Table 4.4. Tolerance of finished bedding plants to cool night temperatures.

Group A (Temperature of 10–12°C)	Group B (Temperature of 14–16°C)
Ageratum	Begonia
Antirrhinum	Catharanthus
Calendula	Celosia
Dianthus	Coleus
Lobularia	Impatiens
Perennials (all)	Zinnia
Petunia	
Phlox	
Salvia	
Tagetes	
Torenia	
Viola	

Fig. 4.2. A greenhouse with the ability to cool plants naturally for toning (Ivy Acres, Calverton, NY).

respiration and a subsequent build-up of carbohydrates essential for stress tolerance. Those in group B are more tender, thus lowering temperatures to 10°C for a prolonged time will do more harm than good. Many of the species in group A or B are also grouped together in the plug storage table (Table 4.1). Younger plants in the plug stage are capable of being stored for longer periods of time and at cooler temperatures than the mature, finished plant.

Lowering greenhouse temperatures may inhibit flowering or growth in immature plants in the same greenhouse. Some greenhouses take advantage of natural cool outdoor spring conditions where trays may be placed during the day and returned to the greenhouse in the evening (Fig. 4.2).

HARDENING OFF WITH FERTILIZATION

Some work has been done with bedding plant plugs (Kaczperski and Armitage, 1993) as well as with finished bedding plant crops to determine the effect of the fertility programme on shelf-life. Research suggests that fertilizer concentration should not be increased prior to shipping or sale. This is particularly true when plants are to be shipped in non-cooled vehicles in the dark. Giving one last 'shot for the road' does nothing but provide additional soft growth, and a greater potential for salt damage when plants dry out (which they invariably will). Discontinuing fertilizer entirely is also poor practice because of the low nutrient pool in the media. This is exacerbated when soilless media are used. The best practice appears to be the reduction of fertilizer rates near the end of the crop cycle. Reduction of the fertility concentration by one-half when buds are visible is a good rule of thumb for most bedding plant species. The influence of hardening off with fertilizer on petunia is shown in Table 4.5.

In some cases, no differences were noted between plants maintained at 200 ppm and those reduced to 100 ppm at visible bud, but in no case was the reduction of fertilizer detrimental to shelf-life. Eliminating fertilizer entirely is not recommended, at least for petunias. For plants with lower sensitivity to soluble salts and those with less time between visible bud and open flower (e.g. *Impatiens*), discontinuing fertilizer entirely at visible bud stage would probably

Table 4.5. Effect of fertility toning of petunia at visible bud stage. Data taken 10 days after shipping. (Adapted from Armitage, 1986b.)

Reducing fertility from 200 ppm to	Visual rating (5 = excellent, 1 = poor)
200 (no change)	3.3
100	4.2
0	2.9

have little effect on shelf-life. In general, however, reducing fertilizer concentration by half at the time of visible bud should be practised.

HARDENING OFF WITH LIGHT INTENSITY
Although reducing the light in the production phase of some pot plants and certainly many foliage species is beneficial to shelf-life, little benefit appears to result in bedding plants. Reducing light to bedding plants reduces carbohydrates and is of doubtful benefit, but little work has been done to test this point. Bedding plants to be shipped for more than 3 days in the dark may benefit from reduced light in the greenhouse; however, in general, hardening off bedding plants by reducing light is of questionable value.

HARDENING OFF WITH WATER
The manner with which plants are irrigated can significantly affect not only the time on the bench and quality of the crop but also their postproduction life. Hardening the plant by reducing irrigation frequency near the end of the crop cycle increases shelf-life. Work with petunias showed that plants are dried out (but did not wilt) declined much less rapidly than those watered 'normally' (i.e. as they had been throughout the production cycle) and those kept constantly moist (Table 4.6).

The effects of the watering frequency in the above experiment were studied under simulated warm (20°C constant) and hot (20°C nights, 30°C days) retail areas. In both retail simulations, damage to overwatered plants showed up rapidly during postproduction, although no damage was detected in the greenhouse. These data indicate that plants should be allowed to dry down, but not to the point of obvious, prolonged wilting, at the end of the production cycle. It is difficult to recommend exactly when to start reducing water, but the visible bud stage seems most appropriate. This practice will not slow down anthesis (the process of flower opening is dependent mostly on temperature), yet plants will be more able to adjust to the retail or garden environment.

Table 4.6. Effects of frequency of irrigation on plant persistence in cool and warm environments. (Adapted from Armitage and Kowalski, 1983.)

Frequency of watering	Days of postproduction life	
	Warm area	Hot area
High	9	4
'Normal'	9	6
Low	15	8

Fig. 4.3. Geraniums being cleaned of senescing flowers and leaves prior to shipping.

Final grooming of plants

An important job in the production phase is removal of dead or dying tissue such as leaves, buds and flowers. Grooming plants in this way not only makes them more attractive but also sharply reduces production of ethylene gas by the dying tissue. Ethylene promotes abscission of buds and, if present in sufficient concentrations, results in leaf drop of some species. During the busy shipping and retail sales season, plants are inevitably mishandled and abused, therefore, it is essential that plants are inspected before they leave the greenhouse (Fig. 4.3). This is their final journey and they should go well dressed.

Responsibility of the Retailer

Visiting retail outlets in the spring of the year affords both delight and disgust: delight that people still want to buy plants in spite of rather poor maintenance and displays seen at some establishments; disgust that some merchants still treat plants like furniture. The scene shown in Fig. 4.4 is unfortunately, not particularly difficult to find during the spring. Chain stores in the United States seem to have more than their share of unkempt outlets, but they are surely not alone.

Fig. 4.4. A chain store in which the plants appear to have survived a war. This scene does nothing for the plants, let alone sales.

Fig. 4.5. A chain store in which plants are neatly displayed, properly labelled and well maintained.

Most retailers, especially chain stores, are not trained to know anything about handling bedding plants. In large outlets, bedding plants are seasonal and only a small part of their offerings. As much as growers would like to think that retailers will learn proper handling techniques, it is the responsibility of the grower to provide as much assistance as possible. This may be in the form of removable carts with displays already prepared at the greenhouse. Such displays can advise on suitable species, sun or shade tolerances and watering practices. However, the retailer must take responsibility for basic plant maintenance practices so that the outlet in Fig. 4.5 is more the rule than the exception.

The retailer's responsibilities include purchasing good quality plants and hiring good personnel to properly maintain them. The maintenance includes supplying shade, aeration and irrigation, and manipulating temperatures where possible.

Plant quality

The retailer who pays a little more for better quality plants is repaid in two important ways. First, higher quality material sells more rapidly, thus accelerating turnover and reducing shelf-life problems. Second, higher quality material persists longer on the shelf. Plants previously subjected to undue stress on the production bench are less able to cope with the greater stress of the retail shelf. The retailer who deals with growers who incorporate postproduction practices into their production programmes has fewer shelf-life problems. Benefits to the grower become more obvious as retailers understand the role growers play in the shelf-life of their product. Growers should be able to market their crop more aggressively and ask top price when they state that it was grown under an integrated postproduction programme.

Personnel

In the spring when it is busy there are seldom enough people in retail outlets who know what they are doing. Nothing turns people away faster than poor looking plants and a messy display area. It is important to provide some training to temporary helpers. At the very least these people should know about the basics of watering and cleanliness.

Plants should not be watered without water breakers. This sounds so simple, yet in many outlets a hose is unavailable, let alone a hose breaker. Knocking over plants by using the unbroken stream from the hose results in damage to plant tissue, ethylene production and reduced shelf-life.

Workers must also be aware of how important it is to keep plants and the area around them clean. There must be enough people even during the busiest times to do this. Removal of dead and injured plants, senesced leaves and flowers, and general carnage must be done at least twice a day during the busy season. Not only do people get 'turned off' when plants are dangling from their roots, but high concentrations of ethylene may be produced. A skilled employee is not necessary to do these things but it does take a competent one. This person will be needed at the busiest season only. Regardless of the type of retail outlet, at least one person should know how to water properly and to keep plants groomed. Often, even this minimum standard care is sorely lacking.

Shading

Plants at retail outlets should be shaded from the sun in areas where temperatures are expected to be 20°C or higher, regardless of species (Armitage, 1993). In general, areas that have no shade are considerably warmer and postproduction life is shortened. Shading can be accomplished with shade cloth, fibreglass panels, canvas cloth, or plants may be arranged under shade trees. Nelson (1984) showed that optimal light levels for most bedding plant species in the retail area are relatively low. Since not all days are equally cloudless and since full sun occurs only between 12 and 3 p.m., 50–80% shade material is sufficient to reduce sunlight to appropriate levels in warm, sunny climates (Armitage, 1993). This is especially true as the season progresses and light intensity increases. Shading also reduces damage from driving rain and other inclement weather. If afternoon temperatures are not expected to rise above 20°C, then the value of shade is minimized.

Ventilation

Lack of air movement due to buildings, fences, traffic or parked cars must be avoided for two obvious reasons. First, air movement is a natural diluter of pollutant build-up. If plants are subjected to ethylene from damaged tissue or vehicles, air movement reduces the concentration, much reducing damage. The presence of vehicle traffic is unavoidable in and near many retail locations and pollutants often rise to dangerous concentrations. Air samples taken from retail outlets in the spring yielded values as high as 0.5 ppm ethylene (Armitage, 1982), which, if sustained over a period of a few hours, can result in damage to sensitive species.

Some plant species as well as certain plant parts are more susceptible to ethylene damage than others (Reid, 1985). Members of the Asteraceae family (e.g. marigolds, zinnia) are little affected by ethylene whereas impatiens, geranium and salvia are susceptible (Table 4.7). Mature flower buds are much more prone to senescence due to ethylene than young flower buds or leaves (Reid, 1985).

Insufficient research has been conducted to determine sensitivity of all bedding plants to ethylene; however, Table 4.7 lists a few sensitivities.

In practical terms, little can be done about vehicle traffic except displaying

Table 4.7. Sensitivity of various bedding plants to ethylene. Based on exposure of 1 ppm for 3, 6, 12, 24 or 48 hours. (Adapted from Reid, 1985.)

Insensitive	Moderate	High
Calendula	Begonia (fibrous and tuberous)	Antirrhinum
Capsicum	Coleus	Impatiens
Lobularia	Petunia	Pelargonium (petal shatter)
Tagetes		Salvia
Zinnia		

less susceptible species, such as zinnia or other members of Asteraceae, nearer the road and those more susceptible, such as impatiens, as far from the source of contamination as possible. Building a display area that does not restrict the flow of air also helps reduce the problem of ethylene.

The second, and probably more important influence of ventilation is that it reduces temperature. Good ventilation, combined with reduced light intensity, helps keep temperature within safe limits. It is not difficult to increase aeration and air movement by using overhead fans in the greenhouse or retail area.

Temperature

This is the most important factor to be controlled in the retail area as it is in the greenhouse and shipping area. If temperatures are maintained near 10°C, then light intensity, fertilizer, shade and ventilation are of little importance. Experiments with bedding plants have shown that cool temperatures are the most important environmental factor for increasing shelf-life. When different production practices such as high or low frequencies of irrigation, different fertilizer regimens or different greenhouse temperatures were used, plants held at 10°C, unlike those in warmer retail areas, showed few differences due to production or shipping treatments (Armitage, 1983; Armitage and Kowalski, 1983; Nelson, 1984). Although most crops such as petunias, geraniums and marigolds persist longer at cool temperatures, others such as impatiens prefer warmer temperatures near 15°C, but only coleus performed best above 20°C (Nelson and Carlson, 1987). Regardless of whether the optimum temperature is 10 or 15°C, obvious practical problems occur.

Maintaining temperatures at 10°C may be pleasant for plants, but not terribly pleasant for people. It is also impractical during the spring and almost impossible during the summer in most areas of the world. The aforementioned factors of shading and air movement help greatly in keeping temperatures reasonable and are, along with good grooming and housecleaning, the best means to ensure high quality plants on the retail shelf.

Retailers can help maintain fresh looking plants by providing well-trained staff, reasonable care and the proper environment. The greatest problems in the postproduction chain occur in the retail area. Fortunately, they are also the easiest to solve. Unfortunately, they will not be solved until growers and retail management are committed to sound postproduction practices and cooperation.

DISEASES AND PESTS

CONTROL OF DISEASES AND PESTS

Control of pests and pathogens is a continuous activity in a production greenhouse. Today, the use of chemical controls is diminishing as pests evolve resistance and alternative methods become more environmentally attractive. Sanitation was and always will be the most important step in the control of insects and diseases. However, as our knowledge base expands, the use of integrated pest management (IPM) systems continue to increase. While IPM systems make great economic and environmental sense, they have been slow to be implemented on a worldwide basis. The blind reliance on pesticides often

Table 5.1. A common sense approach to integrated pest management.

Step no.	Action
1.	Weed control in and around the greenhouse
2.	Sanitation, including washing and pasteurization prior to crop establishment
3.	Inspection and grooming of newly acquired plants
4.	Placing screens (doors, vents) over greenhouse entrances
5.	Routine surveillance, quantification and recording of pest appearances in the greenhouse
6.	Adjustment of the greenhouse environment to inhibit pest but not plant growth
7.	Pest eradication methods of either biological control or pesticide application

reduced the grower to nothing more than a 'spray and pray' gunman.

IPM is based on the integration of preventative, surveillance and corrective measures into a holistic programme for pest control. Total eradication of insects, mites, nematodes, disease organisms and weeds is not the intent, rather IPM is a system in which small populations of pests are tolerated, and chemical controls are minimized. Nelson (1991) outlines the steps in the process, in the order they are likely to be accomplished (Table 5.1).

Weed control

Weeds are a host for many insects and disease organisms and must be controlled. Not only are they potential sources of insects and pathogens, they are also unsightly and project a sloppy image of the greenhouse business. Controlling weeds in and around the greenhouse is necessary for any IPM programme. Only herbicides registered for use in greenhouses may be applied and even these must be used with caution. All herbicides utilized in the greenhouse are best used when no crops are present. If grass is grown around the greenhouse, it should be mowed weekly to a height of 1.5 cm.

Sanitation

Sanitation is the first line of prevention against pests. Pasteurizing soils, and cleaning benches, watering systems and tools with sterilants such as hypochlorite (bleach) should be scheduled on a regular basis. Yellowing or fallen leaves, spent flowers or other plant debris should be discarded, root media removed from benches and floor drains cleaned. Sanitation is a state of mind and should be as routine in the management of the greenhouse as watering is in the management of the plants.

Plant entry

All seedlings and plants brought into the greenhouse must be inspected for pests and diseases. If either are present, the plants must be isolated, treated or discarded.

Insect screens

Polyethylene screens with 400 mesh can be installed over cooler pads and vents to prevent the entrance of many insects. The use of screens over cooler pads requires up to five times more pad surface area but can keep out devastating vectors such as western flower thrips.

Pest surveillance

The entire greenhouse range should be inspected at least twice a week. Insects inevitably prefer certain locations in the greenhouse, particular crops or specific areas. The undersides of leaves must be examined and damage patterns in the greenhouse recorded. Approximately three plants per bench should be routinely inspected and the presence and rate of increase of insects on flagged plants should be recorded. Yellow sticky cards have proved useful for pest surveillance. They are placed over the plants and raised as the crop grows. For small greenhouses, one card per 93 m^2 (1000 ft^2) is sufficient. For large houses, one card per 930 m^2 (10000 ft^2) is a minimum. Cards must be inspected for flying stages of insects and cleaned or replaced each week. Keeping cards in service for weeks or months looks good but is of no use as one cannot tell new entries from old. If the types, numbers, stage of development and rate of development of insects are recorded, then biological or pesticide control will be more effective. In the case of pesticides, monitoring pests results in the proper pesticide being used and applied where and when it will be most effective. Less pesticide is used and the most vulnerable stage of insect is treated with the appropriate chemical.

Environmental adjustments

Many pathogens require high humidity and/or free water on foliage. Heating of the greenhouse prior to dusk and venting to allow drier air into the greenhouse reduces humidity. Similarly, some pests such as spider mites prefer warm, dry areas and monitoring may indicate areas of the greenhouse where use of a cooling system and intermittent syringing of plants may be effective.

Chemicals and biological control

Pests will never be eradicated, however, control of pests to acceptable levels may be accomplished through an IPM programme incorporating registered pesticides and biological controls. Effective pesticides are becoming more scarce as pests become tolerant to many chemicals and environmental agencies continue to reduce the number of chemicals available to the grower. The introduction of new greenhouse pesticides has slowed to a trickle because the relatively small size of the greenhouse industry compared to agronomic crops makes the expense of registering new chemicals prohibitive. The IPM approach does not eliminate pesticides but limits their use to more strategic situations. Growers must learn not only to monitor insect populations religiously, but must know the developmental stages and feeding habits as well. IPM calls for a heavy commitment of time and patience; however an effective IPM programme bears significant payoff in reduced pesticide use,

Table 5.2. Beneficial organisms useful for biological control in the greenhouse (after Nelson, 1991).

Control for	Beneficial organism	Comments	Environment
Aphids	Midge, *Aphidoletes aphidimyza*	1−2 pupa m^{-2}. Introduce 3−4 times every 1−2 weeks	23−25°C, RH 80−90%
Leaf miner	Black wasp, *Diglyphus isaea* or *Dacnusa sibirica*		22−27°C
	Nematode, *Strinernema feltiae*	Spray on plants at night. Allow several hours of drying time	Approximately 21°C
Mealybug	Australian lady beetle, *Cryptolaemus montrouzieri*	1−2 predators m^{-2}. Control ants	22−25°C, RH 70−80%
Mites	Predator mites, *Phytoseiulus persimilis*	24 predators m^{-2}. Reintroduce at 2−4 week intervals	20−27°C, RH 60−90%
	Predator mite, *Phytoseiulus longipes*	As above	25−38°C
	Predator mite, *Amblyseius californicus*	As above	25−32°C, RH >60%
Thrips	Predator mite, *Amblyseius cucumeris*	30 mites per plant	Optimum temp. of 30°C, high humidity
Greenhouse whitefly	Wasp, *Encarsia formosa*	Hang pupae over plants	23−27°C, RH 50−70%, light intensity >7000 lx
Caterpillars and moths	Bacterium, *Bacillus thuringiensis*	Available as a spray under many names	

pesticide resistance, crop injury and possible litigation. Reduced pesticide use also results in happier workers, a difficult to measure benefit but a very important one.

Control of pest and disease organisms through the action of other living organisms is referred to as biological control. Biological control requires strict attention to detail and involves more than releasing beneficial insects. If pest populations are high, they should be reduced with safe pesticides with little or no residual killing power. The number of beneficial organisms released must be in proportion to the population of the pests to be controlled. A strict schedule of monitoring pest density is essential for biological control to be effective. The three types of beneficial organisms are predators, parasites and pathogens. Predators may attack more than one insect and include common garden organisms such as ladybugs, mantis and predatory mites. Parasites are host specific and usually complete their life cycle in or on the host. Parasites of leaf miner are becoming more common as a greenhouse control. Pathogens are microorganisms which cause disease, and eventual death, to their host. The bacterium, *Bacillus thuringiensis*, is sold under a number of product names throughout the world for the control of worms and caterpillars.

Registered chemicals and IPM programmes using parasitic organisms can be obtained through greenhouse companies, chemical firms, private and public IPM specialists and publications dealing with the specific chemicals and IPM tools available in your country. Some of the beneficial organisms available to control greenhouse insects are listed in Table 5.2.

DISEASE AND PEST ORGANISMS

Diseases

Numerous diseases of bedding plants occur and even the most careful grower must be concerned with outbreak of infection. Three main causal agents are involved in pathological diseases in bedding plants. They are viruses, bacteria and fungi, although nematodes can occasionally be a problem. Much of the following information is cited from Mooreman (1985) and Daughtrey and Horst (1990).

Viruses

Viruses are submicroscopic agents composed of protein surrounding RNA or DNA. The most common symptom on the host plant is dwarfing or stunting, and leaves and flowers may be faded, streaked, puckered, blotched or deformed. Viruses may be introduced on seeds (although uncommon) or by vectors such as aphids, nematodes or western spotted thrips. They are further spread through wounds or damaged areas on the plant as well as by insects feeding on a clean plant after feeding on an infected one. Purchasing virus-free indexed stock is highly recommended if geraniums are grown from cuttings to reduce the incidence of other viruses such as tomato ring spot, tobacco ring spot and pelargonium flower break.

The most serious greenhouse disease caused by a virus is the tomato spotted wilt virus (TSWV). It has become a major problem in the greenhouse industry due to its wide host range, worldwide distribution and rapid multiplication. Impatiens and New Guinea impatiens are particularly suscep-tible but, except for geraniums, almost all bedding plants may be attacked by TSWV. Chrysanthemum and gloxinia, also highly susceptible by TSWV, should not be grown in the same area as susceptible bedding crops. Thrips can carry the virus from outdoor crops to a clean greenhouse or the virus may be introduced through planting stock, and cuttings from infected stock plants spread it even further. Symptoms vary among species and identification is difficult because many of the symptoms are similar to those of other diseases. Some indications of TSWV infection include necrotic and chlorotic ringspots, veinal distortion, leaf damage, sunken spots, stunting and death. No cure is known and the best control is through destroying infected plants and controlling thrips.

Bacteria

Few bacterial diseases affect bedding plant species; however, such diseases are very difficult to control. The most significant disease is bacterial blight of geranium caused by *Xanthomonas pelargoni*, resulting in spotting or wedge-shaped yellow areas on the foliage. Black dieback and stem cankers occur as the disease progresses, finally causing total destruction. Even the thought of this disease gives geranium growers grey hair and ulcers. *Xanthomonas*-free cuttings may be obtained through propagators with culture index pro-grammes. Seed-propagated geraniums will not initially carry the bacterium but may become infected if grown with vegetative-propagated cultivars. Control of the disease is practised through the use of culture indexed stock, use of individual tube watering systems to avoid splashing of water, and roguing of infected plants. Although a handful of bactericides exist, attention to cleanliness and careful examination of the crop are the only significant controls. Other organisms such as *Pseudomonas* spp. and *Corynebacterium* spp. may occasionally be a problem; the control of these bacteria is similar to that of *Xanthomonas*.

Fungi

Fungi cause the most diseases in bedding plant crops. They are also the easiest to control. Although many diseases have been diagnosed over the years, the

Table 5.3. Common fungal diseases of bedding plant crops. Adapted from Daughtrey and Horst (1990) and Carlson *et al.* (1992).

Fungal diseases	Causal organisms	Symptoms
Damping off	*Rhizoctonia* *Phytophthora* *Pythium*	Usually first seen in germination flats Poor stands of seedlings Plants collapse at soil line Circles of dead and dying plants enlarge daily Threads of fungi visible on dead plants
Root rots	*Pythium* *Rhizoctonia* *Thielaviopsis*	*Pythium* causes soft dark brown to black rot on roots which may extend to stem base *Rhizoctonia* results in brown or tan dry rot. Usually seen at soil line, root seldom affected *Thielaviopsis* causes a drier black stem lesion than other fungi. May also attack older plants. Pansies and vinca particularly susceptible
Sclerotinia crown rot	*Sclerotinia sclerotiorum*	Cottony-white growth on soil or crown of plant Hard, black resting structures (sclerotia) in cottony growth May be found on older plants Spreads rapidly in warm, damp weather
Grey mould	*Botrytis cinerea*	Soft decay of seedlings Fuzzy, grey growth Spores explode like dust when disturbed Starts on injured or dead plant tissue Spreads rapidly in cool, moist conditions Spreads from top downward
Rust	*Puccinia*	Red leaf spots on foliage Red rust on hands of workers Usually starts on lower leaves and underside Rings of spores form as older leaves enlarge Common in geranium, hollyhock, fuschia and snapdragons
Powdery mildew	*Oidium*	White, fuzzy growth on leaf surfaces Rarely seen on seedlings Common in older zinnias and snapdragons

six most significant are listed in Table 5.3.

Numerous chemicals are available to combat fungal infections, however their availability and legality constantly change. Contact local agencies for chemical means. Environmental control is the best means of preventing diseases. Excellent air circulation and ventilation with fresh air are probably the two most important methods to avoid serious disease outbreak. Heating and venting an hour prior to sunset to exchange warm, humid air for cooler, drier air reduces many fungal problems.

ENVIRONMENTAL STRESS-PROMOTING CONDITIONS

Many fungicides work in concert with proper environmental conditions, not in spite of them. Similarly, if the environment is conducive to fungal growth, the fungi may become much more aggressive and cause problems that would not otherwise occur, even if the organisms were present. Table 5.4 suggests how improper control of the environment enhances root and crown rot fungi (Powell and Lindquist, 1992).

Many organisms become serious problems only when plants are stressed.

Table 5.4. Environmental stress-promoting conditions for root and crown rot diseases (Powell and Lindquist, 1992).

Pathogen	Stress-promoting conditions
Fusarium	Wounds caused by insects or handling High ammonium
Phytophthora parasitica	High temperatures (>25°C) Overwatering or poor drainage
Pythium irregulare	High soluble salts Overwatering
Pythium ultimum	Cool temperatures (<20°C) Overwatering or poor drainage
Rhizoctonia	Alternating extremes of wet and dry High temperatures Wound at the plant's crown High soluble salts
Thielaviopsis	High pH Temperature too high or too low High ammonium

Table 5.5. The influence of pH on the severity of black root rot on pansies (Powell, 1992). Plants rated on a scale of 1 to 5, 5 being healthiest.

pH	Control	Inoculated	Inoculated + Benlate treatment
4	3.5	2.2	2.7
5	4.7	3.8	4.5
6	3.5	2.7	3.0
7	3.0	2.5	3.3

Growing conditions for pansy and vinca call for cool and warm greenhouse environments respectively. *Thielaviopsis basicola* (black root rot) is significantly worse on pansies grown in hot greenhouses and on vinca grown too cool, conditions which cause significant plant stress. High fertility levels, especially if the medium dries out occasionally, also result in greater severity of the disease in pansy (Powell, 1992). When crops are produced under less stressful environments, the fungus seldom is a problem. Similarly, if pH is adjusted to 4.5 to 5.0, *Thielaviopsis* is of little concern (Table 5.5).

The importance of pH in disease control is clearly shown in this work. As the pH deviates from 5.0, the severity of the disease increases regardless of whether the plants were inoculated or treated with a fungicide.

Sterilizing containers, benches and soil used in mixes is essential. Cleaning up debris in and around the greenhouse cannot be ignored if a serious disease prevention programme is to be implemented. Removal of spent flowers or damaged foliage must also be practised religiously. Weeds have no place in a greenhouse and should be eradicated before they build up. A quick tour of a poorly maintained greenhouse rapidly reveals debris, weeds, and general lack of concern about cleanliness. It takes less than one minute to determine if the greenhouse operator is serious about disease and pest control. Poorly maintained greenhouses should be avoided like the plague.

Pests

All the common insects of greenhouse crops also find bedding plant species to their liking. The most common pests of bedding plants are aphids, thrips, spider mites, whitefly (greenhouse and sweet potato), fungus gnats, shoreflies and worms. Leaf miners, snails, slugs and scale insects can occasionally become problems but generally they spend their time on other more exotic species. IPM practices and pesticides are useful for control of many insects.

6

MECHANIZATION OF PRODUCTION

The greenhouse industry is keenly competitive within and among countries and it is imperative to automate and mechanize in order to maintain profitability. The wheelbarrow and hose must give way to flat fillers and controlled irrigation techniques. From greenhouse layout to computer control of fertilization, mechanization has swept the production of bedding plants from a highly labour-intensive to a highly automated activity. The many improvements in greenhouse construction, coverings, bench design, seed technology, soil mixing, irrigation and fertility delivery systems, dibblers and plug dislodgers, tray mechanization, to name but a few, have been important in allowing the technology of bedding plant production to mature. However, the development of plug systems has had the single greatest influence on automation in the bedding plant range. The development of the plug tray and automatic seeder along with advanced control sensors to allow for more precise control of greenhouse environment have all been relatively recent innovations in bedding plant production. The advent of plugs resulted mainly because the technology to automate became available.

Plug technology in bedding plants probably started in the United States around 1969 with the development of trays and a drum seeder to singulate seeds. Development was vigorously pursued and the first commercial plug trays, such as the 648 tray, called the waffle, appeared around 1974. Seeders and trays developed along parallel lines, innovation of tray design being dependent on seeder design. Innovative growers embraced the plug concept in the late 1970s and early 1980s and there has been no looking back since. Today's seeders are capable of sowing 35 to 1100 flats per hour and range in price from US$300.00 to 20000. As more growers used plug trays, seed companies found that high and uniform germination of seed was essential. The improvement in seed technology was the direct result of demand from large growers using plug systems. Other advances in greenhouse mechanization and automation have developed parallel to the growth in plugs.

Greenhouse Layout

The greenhouse layout for bedding plant production should reflect the desire to mechanize the operation wherever possible and allow maximum transmission of winter light. Single greenhouses in areas above 40°N or 40°S latitude should have the ridges running east to west to reduce the effect of shadows; below 40°N or 40°S, greenhouses should be oriented north–south. All attached greenhouses (ridge and furrow) should be oriented north–south, regardless of latitude.

A floor plan must allow for further expansion. The greenhouse range should be built on level land whenever possible to simplify movement of trays or pots. If a slope of the land occurs, construction should begin at the point of average elevation, so that soil excavated may be used as fill in other areas. Most efficient designs incorporate a single, central, service building and houses are constructed around it to minimize distances plants and materials must be transported. Floor plans must consider movement of materials in an efficient manner. One of the costliest phases of traditional bedding plant production is the continual movement of flats or pots from one area to another, often being carried two or three at a time by the employees.

Movement of Plants

Tray (or flat) mechanization allows a low-cost way to move flats around the greenhouse from headhouse to greenhouse and back again. Trays may be moved by rolling aluminium trays (benches) which roll on rails or pipes. A typical tray measures about 1.2–1.4 m by 2.2–3.4 m and is about 10 cm deep. When properly designed, movement of the benches, even when fully loaded, is effortless. The benches may be moved from the transplanting area in the headhouse through a central aisle and into the various bays of the range. One of the main advantages of the moving bench system (Fig. 6.1) is the virtual elimination of aisle space. It is easier to design a new house around a tray system than redesign an old one; however, old houses have been successfully retrofitted.

Self-powered moving belts are also used to move flats or pots of bedding plants from one area to another (Fig. 6.2). They are available in various lengths and widths and can be used to move almost anything from loading or potting areas to their next destination. They can be folded to a small unit for storage or for movement from house to house. They require far less capital investment than movable trays, although they may not be as efficient for bedding plants.

Carts, monorails and metal-wheeled racks have been designed to carry trays of bedding plants through the greenhouse and to fit into trucks. They can even be used as selling units in the retail area, to be picked up and returned to the greenhouse on the next delivery.

Fig. 6.1. Movable bench system.

Fig. 6.2. Conveyor belt system for moving pots.

Irrigation Systems

Irrigation is one of the most labour-intensive and important aspects of bedding plant production. Watering by hand requires significant time and if done improperly results in serious losses to the crops from overwatering, water stress and disease. Automated irrigation is more common in pot plant and cut flower ranges than in bedding plant operations. This is mainly because of the diversity of species and plug, flat and pot sizes used at any one time. Plug growers routinely use fog or fine mist systems based on evapotranspiration or vapour pressure deficits during stages 1 and 2 of plug production. Overhead mist is also common on stages 3 and 4 plugs but manual hose watering is also used. For larger plants growing in flats, overhead watering becomes less desirable due to the possibility of disease spores being spread by splashing water. Plants in pots may be irrigated with tubes, flood bench systems or capillary mats used for other pot crops. Sensors which compute evapotranspiration or vapour pressure deficits are connected to computers which activate the watering system.

Fertilizer can be included in the irrigation stream by fertilizer injectors and piping to carry the solutions to all parts of the range. Each house may be fitted with quick connect clips to the water system so that fertilizer may be applied whenever scheduled. For plug growers, fungicides and growth regulators may also be applied through the water system. These systems lend themselves well to the plug system where hundreds of trays of similar age, size

Fig. 6.3. Automatic sensing and adjustment of conductivity of irrigation water and hence fertilizer concentration.

and cultivar may be treated together. They are not effective when different plant maturities, different container sizes or cultivars are growing together on the same bench. Equipment to monitor conductivity automatically in the irrigation water is also available (Fig. 6.3). If the conductivity is too low, additional fertilizer will be injected to the water supply; if too high, the solution will be diluted. Similar equipment is available to measure pH and perform acid injection for pH control.

Environmental Control

The greenhouse environment is more and more under the control of computers to reduce the labour of turning thermostats on and off, opening ventilators, pulling shade cloth and activating cooling equipment (see Fig. 6.4). Not only are labour costs reduced but the computer or other automated control system often 'babysits' more effectively than do people. This is not to say that people are superfluous in the greenhouse, in fact, people are the most important because they check and control the actions of the babysitter.

Fig. 6.4. Computer controls for the greenhouse.

Production Lines

Flat fillers and pot fillers are becoming more common as the price of labour continues to rise. They may be attached to a system where the flat is filled with media from an automated soil mixer, travels down a belt to be levelled, dibbled and watered. The flat may go to automatic seeder or to the transplant line. Transplants are generally handled by workers positioned at either side of the belt. After sowing or transplanting, flats (pots) may go through an automated label inserter, then be loaded to the movable tray or belt and transferred to the greenhouse. Such a system greatly reduces labour costs and may be paid for with higher efficiency and greater production volume. However, mechanization is not without problems. Initial capital investment is high and breakdowns in equipment are not uncommon. Nevertheless, automation is here to stay and even small growers can take advantage of many of the innovations available.

Transplanters

Plug transplanters are starting to make an appearance, most of which use state of the art computer vision and automation and robotic technology. All models use some form of mechanical 'fingers' which are usually computer controlled. The more expensive models are capable of identifying and rejecting unacceptable plugs. Other options available are the ability to check for empty cells in the final tray, speed and flexibility. Most models are not for small growers of plugs; the speed of transplanters ranges from 3100 to 26000 plugs per hour. Some machines only work with a few sizes of plug and finishing trays while others are sufficiently flexible to use almost any size tray on the market. The cost today ranges from about US$20000 to 80000 (Onofrey, 1993). Research on transplanters continues for simplicity and usability by smaller growers. As demand increases, cost will no doubt decrease.

7

THE FUTURE

The past 10 years has seen continued demand for bedding plants throughout the western world. Breeders from Europe, Japan and the United States continue to improve the habit, colour and performance of many species of bedding plants. Although some years are richer in plant improvements than others, improvements will continue.

Seed technology, including pretreatments, priming, coating, seed enhancing and biotechnology will be an expanding area of research. The advent of expensive equipment to sow seed and transplant seedlings along with the concomitant cost of greenhouse control systems demand vigorous and uniform seed germination and development. This area will continue to take on greater and greater importance in the future.

New species will probably become more important to the floriculture industry. A disturbing trend in grower surveys is the very high dependence on three or four species. While there appears to be more diversity in crops of commercial importance in other countries, the North American market has become fixed on very few plants. Once one has considered impatiens, geranium and petunia, other crops take on a minor role. Even well-known species such as marigolds, vinca and begonias are dwarfed before the big three just mentioned. As recently as 1982, tomatoes accounted for 16% of bedding plant sales in the United States; today they account for about 2%. The market dictates species selection. However, impatiens, geraniums and petunias, no matter how improved, may lose popularity as they become more pedestrian. Additional breeding of new cultivars of vinca, begonia, marigolds, alyssum and salvia will provide a broader base in the future. Marketing of lesser known annuals such as *Nicotiana*, *Mimulus*, *Nemesia* and dozens of other species must also be taken more seriously.

The need for better and more productive marketing of bedding plants is necessary. While the sales of plants today are relatively healthy, many more could be sold, regardless of country, with more effective marketing. Marketing is the key to success of any business and national programmes should be supported by all segments of floriculture to spread the message about the beauty the plants bring to the world. Bedding plant grower organizations exist in most producing countries and should be actively supported. It is not too great a jump to envisage global marketing of plugs and seedlings, although today's quarantine and agriculture laws preclude such importation to many countries. Densely packed units such as plug trays, however, may economically be moved within countries if politically and economically feasible.

Greenhouse computerization and advances in environmental sensors must be continued so that optimum environments for various stages of growth can be realized. The ability to take advantage of stage 1 or 2 metabolism or to use DIF as a regulatory tool is limited by the degree of environmental control in the greenhouse. Better, more reliable environmental sensors are an absolute must for optimizing production techniques. Less technical, but nevertheless providing more flexibility, is the move towards movable indoor/outdoor benching systems. The plants may be grown cool and hardened off prior to shipment. The concept is as old as horticulture, but the system requires significant engineering and growing knowledge.

Local markets will stress more regional plant species and cultivars as more information on the outdoor performance of various cultivars becomes known. Already, autumn sales are limited to southerly areas of moderate winter temperatures. Rain, heat, cold and cloud cover should dictate those plants in the local market, not national or international trends.

General greenhouse concerns for any crop are also important to the future of bedding plants. More reliance on integrated pest management, water recirculation, reduction of fertilizer runoff and, in general, more environmental consciousness will be far more apparent in the future if the industry is to grow. Interdependence between industry and organizations committed to research must become stronger. Governments will no longer bear the cost of ornamental research, and if individual firms do not commit a small portion of their budget to supporting research through check-off programmes or grants, the pace of progress will undoubtedly decline.

The future for bedding plant production and the use of these plants outdoors is enormous. The recent past has seen an unprecedented explosion of information resulting in undreamed of technology and market demand. The future, provided quality of product and integrity of service are stressed, should be outstanding for the grower and end-user of bedding plants.

APPENDIX:
GUIDELINES FOR COMMERCIAL PRODUCTION OF 15 BEDDING PLANT CROPS

The following information is provided to help the reader understand the steps involved in the commercial production of some of the more important bedding species. The information has been gathered from research papers, growers and seed company publications and lectures. Cultivars listed may or may not be available in various countries or may be available under different names. The notes are guidelines only and will differ according to availability of materials and the ability to maintain greenhouse environments in different locations.

ANTIRRHINUM MAJUS

Snapdragon
Scrophulariaceae

Snapdragons are grown as cool season annuals and tolerate cool greenhouse conditions. They perform best outdoors in areas of cool summers and mild winters. In warm areas of the world, they may be autumn planted and overwintered for early spring flowering. They are also extensively grown as a cut flower crop in the field and greenhouse.

Forms available

Snapdragons are categorized by height and range from dwarf (15–30 cm), medium (30–60 cm) to tall (>60 cm). The first two forms are used as bedding plants, the latter as cut flowers.

Cultivars sold as bedding plants[a]

Dwarf: Bells series, Chimes series, Floral Carpet series, Floral Showers series, Pixie mix, Princess series, Royal Carpet mix, Tahiti series, Sunshine mix.
Medium: Butterfly mix, Liberty series, Sonnet series.
Tall: Rocket series.
a = many other cultivars are sold for greenhouse cut flowers.

Propagation

By seed; pre-chilling for about a week at 5°C enhances germination. One seed per cell in plug tray. Seeds require light for best germination, therefore seeds should be sown on surface.
Temperature: 21–24°C, germination begins in 4–8 days, completed by 14 days. Avoid water stress.
Light: Light should be provided if germinating in growth rooms. Do not cover seed deeply.

Growing-on

Transplanting
– *Traditional*: Grow seedlings at 15–17°C until transplant (two to four leaf stage).
– *Plugs*: 6–7 weeks before transplanting to final containers.
Temperature: Grow transplants (traditional) or stage 2–4 plugs at 15–17°C. After transplanting, temperature should be lowered to 8–10°C. Warmer temperatures cause plant stretch. In spring, flats and finishing pots may be moved outdoors after the threat of frost has passed.
Photoperiod: Long days promote flowering although natural days are commonly used in a commercial greenhouse.
Light: Supplemental light is useful (50–75 μmol m^{-2} s^{-1}) for 14–18 hours per day.
Fertilization: No more than 100 ppm N, especially when temperatures are below 15°C. Plants are susceptible to soluble salt damage. Maintain pH 5.5–5.8, soluble salts approx. 0.75 mmho cm^{-1}. High pH causes interveinal chlorosis due to tie-up of iron. High ammonium and high salts are seen as dark green leaf tips.
Height control: Should not be necessary if grown cool. If cool temperatures cannot be maintained, paclobutrazol (2–5 ppm) or ancymidol (100 ppm) may be used. DIF is also effective.

Common problems

Physiological: Susceptible to high salts, particularly nitrogen. Boron is often lacking in artificial media and should be applied early in the crop cycle.

Pests: Western flower thrips cause damage and also transmit tomato spotted wilt virus (TSWV), aphids, whiteflies and leaf miners.

Diseases: Damping off (*Rhizoctonia, Pythium*) and rust are serious problems. Both are more prevalent under conditions of high soluble salts. Rust is more of a problem in the field than the greenhouse. TSWV is becoming more prevalent as thrips become more of a problem.

Postproduction concerns
Reduce fertilizer and temperature after flowering to harden off plants.

Schedule
Traditional: Tall cultivars are sold without flower in 7–9 weeks after sowing; dwarf and medium forms 8–11 weeks in flower depending on temperature used.
Plugs: 6–7 weeks to transplant, then an additional 5–7 weeks for flats or 7–8 weeks for 10 cm pots.

BEGONIA SEMPERFLORENS-CULTORUM

Wax begonia, Fibrous begonia
Begoniaceae

The limitation to growing wax begonias has always been their long greenhouse production time, requiring up to 20 weeks for older cultivars, and difficult seed propagation. Recent research has enhanced seed germination and reduced growing time significantly. New cultivars provide flowers in white, red and pink and are some of the most carefree plants available to the gardener and landscaper. Plants are tolerant of sun and shade in the garden.

Forms available
Wax begonias are sold as dwarf (12–20 cm), intermediate (20–25 cm) and tall (>25 cm) with green and bronze foliage. Hanging basket forms have also been bred.

Cultivars
Dwarf: Ambassador series, Atlanta series, Cocktail series[a], Coco series[a], Hot Fudge series[a], Olympia series, Rio series[a], Rusher series, Varsity series.

Intermediate: 'Othello'[a], Party series, 'Scarlanda', 'Scarletta', 'Viva'.
Tall: Danica series[a], Frilly series, 'Glamour', Lotto series, Sensation series.
Hanging basket: Avalanche series.
a = bronze foliage.

Propagation

Propagated by seeds which are often pelleted for easy handling. Seeds must be sown in a light, well-drained medium and should not be covered. Two seeds are often sown per plug. Fine mist or fog is essential to ensure that seeds are not washed away.

Temperature: Air temperature 23–27°C, soil temperature 21–23°C. Germination begins in 5–7 days, and occurs over 14–20 days. Red light reduces germination time significantly.

In plugs, begin stage 1 at 25–27°C, and after 6–7 days reduce temperature to 22–26°C for stage 2 (21 days).

Light: Light is necessary for germination. Light should be applied for 18 hours if germinating in chambers.

Growing-on

Transplanting

– *Traditional*: Transplant when seedlings are easily handled (one or two true leaves), place in shaded area overnight to reduce transplant shock.

– *Plugs*: Plants remain 8–9 weeks in the plug before transplanting to the final container.

Temperature: In the traditional method, transplants are grown-on at 21°C. For plugs, stage 3 may be grown at 21–24°C and lowered at stage 4 to around 20°C. Warmer temperatures (23°C) cause faster growth and flowering but result in some internode elongation in certain cultivars. Plants may be placed under cool temperatures (cold frames, etc.) after temperatures warm up and although foliage colour of bronze cultivars is enhanced, flowering time will be significantly increased.

Photoperiod: Long days (>12 hours) accelerate growth but plants flower regardless of photoperiod.

Light: Begonias are high light plants in the greenhouse; low light results in tall plants in the plug or flat.

Carbon dioxide: Early application of approximately 1000 ppm CO_2 significantly reduces the time to transplantable size.

Fertilization: In the seedling stage, 25–50 ppm N of KNO_3 may be applied; increase to 100–125 ppm N of a balanced complete fertilizer by stage 4 of plug growth. No more than 125 ppm N is necessary at any time in the crop cycle if a constant liquid feed programme is used. A pH of 5.5–5.8 and 0.75 mmho cm^{-1} soluble salts (1:5 dilution) is recommended. High pH (>6.8) and salts below 0.50 and greater than 1.5 mmho cm^{-1} cause stunted growth.

Common problems
Pests: Aphids and mealybugs.
Diseases: Damping off (*Pythium* and *Rhizoctonia*) and grey mould (*Botrytis*) are the main problems.

Postproduction concerns
Plants are brittle and must be handled with care. Fertility and water should be reduced when the flower buds are visible.

Schedule
Traditional: 15–17 weeks from seed to sale.
Plugs: 8–9 weeks in the plug, then 6 weeks for flats, 8–9 weeks for 10 cm pot.

BEGONIA × TUBERHYBRIDA

Tuberous begonia
Begoniaceae

This crop is particularly important for hanging baskets but new breeding has also made this a popular item for pot plants and for landscapers in areas of cool summers. These begonias do not tolerate hot, humid conditions and have limited uses in such climates.

Forms available
Large and small-flowered forms, single and double flowers as well as green and bronze-leaved forms have been bred.

Cultivars
Fortune series, Galaxy series, Memory mix (large-flowered, green, double), Musical series (small-flowered, green, double), Nonstop series (medium flowers, green, double), Pinup series (large-flowered, bronze, single).

Propagation
Although tubers are available to homeowners and pot plant growers, most tuberous begonias are propagated from pelleted seed. Seeds must be kept consistently moist (90–95% RH) and should not be covered.

Temperature: Air temperature 23–27°C, soil temperature 20–23°C. Germination occurs in 10 to 15 days and is complete by 30 days. Cool temperatures delay germination.

In plugs, begin stage 1 at 24–26°C (7–10 days); reduce to 21–22°C for stage 2 (about 3 weeks).

Light: Light is necessary for germination, supplemental light (18–24 hours) enhances germination and early growth.

Growing-on
Transplanting
– *Traditional*: Seedlings are transplanted when the third true leaf is about 1 cm wide (7–10 weeks from sowing), or when they can be handled without damage.
– *Plugs*: Plugs are grown for 9–10 weeks before transplanting to final containers.

Temperature: In the traditional method, seedlings are grown-on at 22–23°C days and 17–18°C nights. In the plug method, stage 3 plugs are grown around 20°C for 3 weeks, followed 2 weeks at approximately 17°C. Cool temperatures (<17°C) cause tuber development at the expense of foliar development. After flower buds are visible, temperatures may be lowered to 13–15°C.

Photoperiod: Tuberous begonias are long day plants for flowering and top growth and should receive long days immediately after emergence until flower buds are visible. Daylength (18–24 hours) or night break lighting (2200 to 0200 h) is effective.

Light: During winter and early spring, supplemental lighting of 60–150 µmol $m^{-2} s^{-1}$ for 18 hours per day is effective.

Carbon dioxide: 1000–1500 ppm during the day accelerates growth.

Fertilization: In the seedling stage, 25–50 ppm N of KNO_3 is effective, followed by about 100 ppm N constant liquid feed of a complete fertilizer. Approximately 250 ppm N per week is recommended. Excess nitrogen causes excess foliage.

Height control: Usually not necessary, if grown properly. However, daminozide (2500–3000 ppm) and chlormequat (500–1000 ppm) can be used.

Common problems
Physiological: Low humidity during germination causes seedling death in the plug or seed tray. Low temperatures or short days cause tuber formation, causing stunted growth.

Pests: Thrips, cyclamen mite, whiteflies, mealybugs and aphids.

Diseases: Damping off, powdery mildew and grey mould are all possible on tuberous begonias.

Postproduction concerns
Most plants are sold in 10–15 cm pots and are relatively large compared with other bedding plants. They are brittle and must be handled with care. Flowers tend to abscise rapidly (shatter) during shipping. 50 ppm of silver thiosulfate (STS) applied to the top of the plants prior to shipping reduces shattering.

Schedule
Traditional: 18–24 weeks from sowing to a 10–12 cm pot. Baskets may require 2–3 weeks longer.
Plugs: 9–10 weeks in the plugs, then 7–9 weeks for 10–12 cm pots, 9–11 weeks for baskets.

CATHARANTHUS ROSEUS

Madagascar periwinkle, Vinca
Apocynaceae

Plants have been grown as pot plants and bedding plants. Some excellent breeding has resulted in large-flowered plants with pink, white, lavender and bicolour flowers. They require warmer temperatures than most other bedding species and are most successfully produced in a separate greenhouse. Vinca are more economically produced by growers in warm climates.

Cultivars
Cooler series, Little series, Pacifica series, 'Parasol', Pretty series and Tropicana series are most popular.

Propagation
Always propagated from seed; cleaned and graded seeds are often available.
Temperature: Germination is best at 24–27°C although some growers germinate at 32–35°C for 3 days, then lower temperatures to 24–27°C. Germination begins in 4 to 6 days and is generally completed in 10 to 15 days.
Light: Light is recommended for germination, seeds should be covered lightly.

Growing-on

Transplanting
– *Traditional*: Seedlings are transplanted at the four-leaf stage, approximately 4–5 weeks from sowing.
– *Plugs*: Plugs are transplanted about 6–7 weeks after sowing.
Temperature: In the traditional method, seedlings are grown at 18–24°C until approximately 2 weeks prior to sale. At that time, temperatures may be lowered to about 18°C. In plugs, 20–22°C is applied in stage 3, followed by 18–20°C during stage 4.

Low temperatures (<18°C) cause yellowing of foliage, stunted growth and a poor quality crop. Low temperature is the most common cause of crop failure with this species.
Photoperiod: Not significant.
Light: Vinca is a high light plant.
Fertilization: Approximately 100–200 ppm N from a balanced fertilizer is sufficient. Fertilization should occur even at cool temperatures. A pH of 5.5–6.0 is optimum. pH above 7.0 causes iron chlorosis. Magnesium should be applied as a chelate to reduce the incidence of magnesium deficiency. Use of nitrogen fertilizers high in ammoniacal nitrogen should be avoided.
Height control: Daminozide (2500 ppm) is often used and is particularly effective when applied during stage 3 of plug growth. Ancymidol and paclobutrazol are effective but both may cause foliar damage if applied incorrectly. DIF is effective as long as temperatures are not lowered below 18°C for more than a few hours.

Common problems

Physiological: Cool temperatures and overwatering in the greenhouse and in the landscape are the most common reasons for lack of success. Irrigation outdoors is best accomplished by drip irrigation rather than overhead sprinklers.
Pests: None of significance, although aphids and thrips can occasionally be problems.
Diseases: *Thielaviopsis*, *Pythium* and *Rhizoctonia* are most prevalent when plants are stressed, usually as a result of cool temperatures (<18°C). *Pythium* produces black lesions on the root and the medium has a musty smell. Roots are generally attacked from the ends. *Thielaviopsis* produces similar lesions without the disagreeable smell and causes older leaves to turn yellow and abscise.

Postproduction concerns

Plants should be marketed under enclosed structures to reduce cold stress during the spring.

Schedule
Traditional: 10–14 weeks from seed to sale.
Plugs: 6–7 weeks in the plug flat, then 5–7 weeks for flats, 6–8 weeks for 10 cm pots.

CELOSIA ARGENTEA

Cockscomb, Celosia
Amaranthaceae

Available in three forms and numerous colours, celosia is popular as a bedding plant, pot plant and a cut flower.

Forms available
Crested form: The common cockscomb (*C.a.* var. *cristata*) in which plants produce many flowers in a convoluted inflorescence. Although they are commonly referred to as cockscomb, I think they look more like coloured brains.
Plumose form: The feathered form (*C.a.* var. *plumosa*) has flowers arranged in an upright inflorescence.
Wheat form: Wheat celosia (*C.a.* var. *spicata*) is a tall form with one or two cultivars. Generally used as a cut flower.
    The crested and plumose forms are available as dwarf, medium and tall cultivars.

Cultivars
Crested: Coral Garden mix (dwarf), 'Toreador' (semi-dwarf), 'Fireglow' (semi-dwarf), Jewel Box mix (dwarf), Chief series (tall).
Plumose: 'Apricot Brandy' (medium), Castle series (medium), Century series (tall), 'Forest Fire' (tall), Geisha series (dwarf), Kewpie series (dwarf), Kimono mix (dwarf), 'New Look' (medium), 'Red Glow', 'Sparkler' (tall).
Wheat: 'Flamingo Feather' (tall), 'Flamingo Purple' (tall).

Propagation
Propagated by seed.
Temperature: Seeds are germinated at 24°C, then lowered to approximately

22°C after about 10 days. Germination commences in 4–5 days and is complete by 10 days.

Light: Light benefits germination, therefore seeds are not covered in the plug flat or germination tray.

Growing-on

Transplanting

– *Traditional*: Transplanting occurs at the two to four-leaf stage, around 3 weeks after sowing.

– *Plugs*: 5 to 6 weeks are spent in the plug prior to transplanting to the final container.

Temperature: Night temperatures of 15–18°C combined with day temperatures around 21°C result in continuous growth. Temperatures below 15°C cause plant stunting and stress.

Photoperiod: Celosia will flower, regardless of photoperiod. However, celosia is a quantitative short day plant and flowers prematurely if grown under days less than 14 hours (Piringer and Borthwick, 1961). Long days promote faster growth and taller plants.

Light: Celosia is a high light plant and benefits from clean greenhouses and supplemental light, particularly in the winter and early spring.

Fertilization: During stage 2, 50 ppm of N from KNO_3 is beneficial. Constant liquid feed of 100–150 ppm N is recommended during growing-on. Plants are highly susceptible to excess soluble salts, causing stunted plants and poor quality.

Height control: Use of DIF is effective. Daminozide (2500 ppm, twice) or chlormequat (750–1500 ppm) are also applied as sprays.

Common problems

Physiological: Premature flowering is a problem if plants are stressed or growth is interrupted. This often occurs when plants are grown too cold or with insufficient fertilizer.

Pests: No major pests occur although spider mites and aphids can be serious.

Diseases: Damping off (*Pythium* and *Rhizoctonia*) of seedlings is not uncommon.

Postproduction concerns

Plants should not be allowed to dry out or remain under cold conditions. Leaves become red and plants are slow to regrow when transplanted to the garden.

Schedule

Traditional: A flowering crop requires approximately 10 weeks from seed to sale in packs, another week in a 10 cm pot.

Plugs: 5 to 6 weeks in the plug, followed by 4 to 6 weeks in the final container.

IMPATIENS WALLERIANA

Impatiens, Busy Lizzie
Balsaminaceae

The breeding of new cultivars has caused impatiens to become a leading bedding plant in most countries. Breeders have developed compact and well-branched cultivars in many colours.

Forms available
Flowers of most cultivars are single but vegetatively propagated plants bearing double flowers are also available. Cultivars are also available in green and bronze foliage.

Cultivars
Almost every major flower seed company has developed a series of impatiens. All single-flowered cultivars are F_1 hybrids, doubles are grown either from seed or from terminal cuttings.

Propagation
All single-flowered forms are propagated by seed. Seed counts vary from cultivar to cultivar. Seeds should be sown in a light, well-drained medium and not be covered. pH should be 5.5 to 5.8; below 5.0 inhibits germination.
Temperature: Soil temperature between 21 and 25°C is best for germination. Germination begins in 4–5 days and is completed in 15 days.

In plugs, seeds should be started at 24–27°C for 3 to 5 days (stage 1), then temperatures reduced 22–24°C (stage 2).
Light: Light is necessary for germination. Natural light may be provided in the mist bed in the greenhouse or artificial light may be supplied in the germination room. Twenty-four hours of light from fluorescent or high pressure sources aid germination and initial seedling growth.

Growing-on

Transplanting

– *Traditional*: Transplanting occurs as soon as plants can be handled, usually after two or three true leaves have developed. Transplant as early as possible to avoid transplant shock. Seedlings are easily damaged.

– *Plugs*: 5 to 6 weeks, depending on cultivar and climate.

Temperature: Night temperatures of 15–18°C, day temperatures of 21–24°C produce high quality plants. Lower temperatures delay growth and flower initiation. Temperatures above 27°C cause stretching and leaf necrosis, particularly if light intensities are high.

Photoperiod: No photoperiod response has been shown.

Light: During the winter, no shading is necessary; however, greenhouse shading is recommended as the season progresses.

Fertilization: Impatiens do not require high levels of fertilization. In plug stage 2, application of 75 ppm N per week by KNO_3, which may be raised to 100 ppm N with a complete fertilizer during the remainder of plug growth, is recommended. In the traditional method, 100 ppm N (constant liquid feed) using a complete fertilizer is sufficient. Soluble salts should be monitored and maintained around 0.75 mmho cm^{-1}. Excess soluble salts result in excess foliage, leaf necrosis and floral abortion.

Height control: Use of DIF is effective (18°C days, 20°C nights). Most growth regulators are effective in controlling internode elongation. Daminozide, paclobutrazol and uniconizol are used.

Common problems

Physiological: Low light, high fertility and overwatering result in stretched plants, leafy stems and poor flowering. Impatiens are also very sensitive to ethylene which causes flower bud abortion. Sources of ethylene (vehicles, abscising leaves and damaged plants) should be avoided in areas where impatiens are grown or displayed.

Pests: Aphids, thrips and spider mites are common to impatiens. Distorted and stunted leaves are normal plant responses to most pests.

Diseases: *Botrytis* is a problem when plants are held in the retail area or in the greenhouse. Cool, moist conditions aid the growth of grey mould. *Rhizoctonia* and *Pythium* are fungi responsible for damping off.

Postproduction concerns

Plants should not be hardened-off by cooling in the greenhouse prior to shipping. Cool temperatures during early retailing cause chilling damage in the form of stunted plants, red foliage and flower drop. Plants wilt rapidly and require constant maintenance in the retail area.

Schedule

Traditional: In packs, 8–10 weeks is necessary from seed to flower. For 10 cm pots, 10–11 weeks is sufficient.

Plugs: 5 to 6 weeks are spent in the plug, followed by 4 to 6 weeks in the final container.

IMPATIENS ×
HYBRIDA (I. HAWKERI)

New Guinea impatiens
Balsamainaceae

New Guinea impatiens have recently become popular as bedding plants. Most are still grown as baskets or potted plants, sold for container plants on a patio or for window boxes. The majority of cultivars are propagated from tip cuttings, although a few seed-propagated cultivars are also available.

Cultivars
Flower colours include white, red, pink, orange and bicolours. Foliage is also an important breeding consideration and plants are available with green, bronze or variegated leaves. In the last 10 years, major breeding programmes significantly improved old vegetative cultivars and by now half a dozen series, each with eight to ten colours, have been developed.

Old seed varieties are 'Tango' and 'Sweet Sue', both large-flowered, tall, orange-flowered cultivars. A recent breakthrough in F_1 seed is the Spectra series, a relatively dwarf form popular with bedding plant growers.

Propagation
Most are propagated from cuttings, a few from seed. Unrooted cuttings generally root in perlite in 7–21 days, but rooted cuttings are available from most propagators. Rooting temperatures are 21–24°C. Seed is germinated similarly to bedding impatiens, but germination is seldom as rapid or uniform, requiring a few more days in the germination area.

Growing-on
New Guineas are grown one cutting or seedling per 10 cm pot, three per 25 cm basket or four per 30 cm basket.

Temperature: After planting, plants should be maintained at 18–20°C night temperature and 21–24°C day temperature. Once established, night temperature may be lowered to 15°C, days to 20°C. Warm temperatures (21–24°C) result in rapid growth. Plants must be well watered, especially at warmer temperatures. Temperatures below 15°C should be avoided.

Photoperiod: No significant responses.

Light: New Guinea impatiens tolerate higher light intensities than bedding impatiens and respond well to high light and moderate temperatures. As much light as possible should be provided in winter, spring and early summer but some shading may be necessary in late spring and summer to reduce excessive temperatures.

Carbon dioxide: 1000–1500 ppm CO_2 increases rate of growth and plant quality. Day temperature may be raised 2–3°C if supplemental CO_2 is used.

Fertilization: Plants respond to high levels of nutrition and 200–250 ppm N from a complete fertilizer at each irrigation may be provided. Pale, mottled leaves and poor growth are symptomatic of low fertility.

Height control: None should be necessary.

Common problems

Physiological: Low light and low fertility cause mottled foliage. Water stress causes leaf and bud abscission.

Pests: Spider mites, cyclamen mites, thrips, mealybugs and aphids enjoy this crop.

Diseases: *Pythium*, *Phytophthora* and *Rhizoctonia* cause root rots, especially if plants are overwatered. Tobacco spotted wilt virus (TSWV), often carried by thrips, can be a serious problem. *Botrytis* is also a problem under cool, humid conditions.

Postproduction concerns

Plants should be watered well prior to shipping. Areas or situations where ethylene builds up should be avoided.

Schedule

Pots (10–12 cm) may be produced in 8–10 weeks from rooted cuttings in warm areas, or in 10–16 weeks in areas of low light and cool temperatures. Baskets require an additional 2 weeks.

LOBELIA ERINUS

Lobelia
Campanulaceae

An old-fashioned bedding plant, lobelia is presently grown in flats or hanging baskets. Plants are used to line walkways, provide colour in the front of borders, accents in window boxes or as hanging baskets on patios and porches. They are popular in areas of cool to moderate temperature; they are short-lived in areas with warm humid summers.

Forms available
Compact forms of lobelia have been bred for pots or pack production and cascading forms are used for basket production.

Cultivars
Compact forms: 'Cambridge Blue', 'Crystal Palace', Moon series, Riviera series, 'Rosamond' and 'White Lady' are some of the cultivars on today's market.
Cascading forms: Fountain series, Palace series, 'Hamburgia', Regatta series and 'Sapphire'.

Propagation
Although plants may be produced from terminal cuttings, most commercial production results from seed propagation. If the traditional system is used, seeds are generally direct-sown to the final container; two or three seeds per cell or pot.
Temperature: Germination occurs rapidly at 24–27°C, after which temperatures may be lowered by 4–5°C. Germination should begin in 4–6 days, although non-uniformity may cause germination to occur over a period of 15–20 days.
Light: Light is not necessary for germination; however, the seeds are so small, they should not be covered.

Growing-on
Transplanting
– *Traditional*: Seeds are generally direct-sown; however, if transplanting is necessary, it is done 2–4 weeks after sowing. Clumps of seedlings are transplanted.

– Plugs: Transplanting to final container generally occurs 5–6 weeks after sowing.

Temperature: Once plants have become established in the final container (traditional) or by stage 3 (plug), temperatures should be lowered to 15–17°C, then lowered to 13–16°C nights and 16–18°C days once plants are in their final containers. High temperatures cause stretch of internodes.

Photoperiod: No significant photoperiod response is known for growth or flowering.

Light: Lobelia grows best at cool temperatures and high light intensity.

Fertilization: Plants are light to moderate feeders. Application of 100–150 ppm N with a complete fertilizer at each irrigation is sufficient. High nitrogen delays flowering. pH should be maintained at 5.5–6.5

Height control: In general, height of lobelia is not a problem, unless grown under low light and warm temperatures. Daminozide (3000–5000 ppm) is effective if required. Apply when plants are young, prior to stretch.

Common problems

Physiological: Warm temperatures result in loss of quality. Plants are particularly susceptible to ozone injury, causing leaf stippling and silvering.

Pests: Leaf miners, whiteflies, thrips and aphids attack lobelia.

Diseases: *Alternaria*, often carried by seed, causes damping off of seedlings. Plants are also susceptible to tomato spotted wilt virus (TSWV), carried by thrips. Thrips must be controlled to reduce the incidence of the virus.

Postproduction concerns

Keep from warm areas of the retail or shipping area.

Schedule

Traditional: Approximately 10 weeks from sowing to flower are needed.

Plugs: Seedlings remain in plugs for 5–6 weeks, then require an additional 4–6 weeks for flats or baskets.

PELARGONIUM ×
HORTORUM

Annual geranium
Geraniaceae

Geraniums have long been popular with consumers throughout the world. Most geraniums are propagated vegetatively and produce double flowers. A relatively recent trend in the geranium market is the production of geraniums from seed. This development allowed bedding plant growers to weave geraniums into their normal system of seed propagation instead of using cuttings. Flowers are single and not as full as the double forms from cuttings. This section deals only with those propagated from seed.

Forms available

Seed geraniums are available in shades of pink, white, orange, scarlet, red and bicolours. The foliage often has a zone of anthocyanin in the middle, thus the common name of zonal geraniums. Some doubles have been bred, but cultural problems and seed availability have limited their use.

Cultivars

Over 150 cultivars are available. The main trend in seed geranium breeding has been towards reducing flowering time. New cultivars include those with orange flowers and highly zoned foliage. Bronze foliage seed-propagated cultivars are almost ready for the market.

Propagation

Temperature: Seeds sown at 21–25°C and high humidity will emerge in 3–5 days.
Light: Light is not necessary for germination.

Growing-on

Transplanting
– *Traditional*: Transplant when seedlings may be handled, about 2–3 weeks after sowing. Delay of transplanting delays growth and flowering.
– *Plugs*: Plants remain in plugs for about 6 weeks.
Temperature: Geraniums are generally grown at 17–18°C night temperature and 20–24°C day temperature. Temperatures lower than 15°C significantly delay flower initiation and development. Anthesis may be controlled by adjusting temperature between visible flower bud and anthesis. The warmer the temperature, the faster the development.
Photoperiod: No significant response.
CO_2: reduces time to transplant and flower. Apply 1000 ppm for 3–5 weeks when plants are young.
Light: Geraniums are high light plants and flower initiation is directly related to light intensity. Supplemental light (High Intensity Discharge (HID) sodium or metal halide, 18 hours per day) is highly recommended during the plug stage or for the first 5 weeks after transplanting in the traditional system.
Fertilization: Plugs may be fertilized once stage 2 has been reached. Transplanted seedlings (traditional method) may also be fertilized with a dilute

solution (75 ppm N) of a balanced fertilizer. Use of 150–200 ppm N on a constant liquid feed basis or 300–400 ppm once a week is recommended. A pH of 5.8–6.2 must be maintained; low pH can result in serious iron toxicity problems.

Height control: Height may be controlled with low temperatures and low fertility regimes, although flowering will be delayed. Chlormequat (1500 ppm) and ancymidol (100 ppm) are the most useful chemicals for height control. Chlormequat should be applied twice at 1500 ppm or more often at dilute rates (e.g. four times with 750 ppm). Chlormequat and ancymidol accelerate flowering by 7–10 days.

Common problems

Physiological: Petal shattering is caused by ethylene, produced naturally or as a pollutant. Cool temperatures after first flower anthesis delays shattering. Silver thiosulfate (STS) has been sprayed onto developing flowers to reduce shattering, but plants must be treated with a fungicide to control *Pythium* prior to STS application. High levels of iron can cause stunting of plants and yellowing of foliage. High amounts of iron or macronutrients should be avoided.

Pests: Pests are seldom a problem; however, aphids, spider mites and mealybugs can attack geraniums.

Diseases: *Botrytis* is particularly bad if spent flowers are not removed and if plants are spaced too closely together. *Rhizoctonia* and *Pythium* can cause serious disease, including seed rot, damping off and black leg.

Postproduction concerns

Cold temperatures in the retail area result in red-leaved stunted plants. Areas of high vehicle traffic should be avoided to lessen ethylene build-up and reduce petal shatter.

Schedule

Traditional: Plants grown under 'normal' greenhouse conditions require approximately 12 weeks from seed to flower, but significant differences exist among cultivars. Fast crops (11 weeks) can be produced with fast-flowering cultivars, supplemental light, growth regulators and warm temperatures.

Plugs: Plants remain in plug flats for about 6 weeks, then are transplanted to flats or 10 cm pots. Plants require 6–7 weeks to finish.

PETUNIA × *HYBRIDA*

Petunia
Solanaceae

Some of the most popular bedding plants in the world, petunias have undergone continuous breeding for the last three or four decades. The results have provided petunias in almost all conceivable colours, three horticultural forms and single and double flowers.

Forms available

Grandifloras: Plants have large flowers (up to 10 cm in diameter). These plants have historically been most popular but do not hold up as well under adverse garden conditions. Single and double-flowered cultivars are available.

Multifloras: Plants have more but smaller flowers than the grandifloras. Multifloras, in general, perform better in the garden, therefore, breeding and production has shifted to the multifloras in recent years. Single and double-flowered cultivars are available.

Floribundas: These are recent hybrids between single multifloras and single grandifloras. This category is used but questionable. There is little enough difference between large flowered multifloras and smaller flowered grandiflora cultivars and this class blurs the distinctions even more. It was probably as much designed for promotion and sales as anything.

Cultivars

Because of the importance of petunias in the last 20 years, every major flower breeder has their own cultivars. Most companies have introduced three to six series of multifloras and grandifloras and a few are selling floribundas. Each series contains five to ten colours plus a mixture.

Propagation

Always propagated from seed.

Temperature: Seeds are germinated at 24–26°C. Germination begins in 2–3 days and should be completed within 10 days. Once seedlings are visible, temperatures may be lowered to 22°C.

Light: Seed is small and should be sown on the surface. Light is not necessary for germination but if germinated in the dark, emerging seedlings must be moved to the light immediately. If sowing in growth rooms, light is recommended.

Growing-on
Transplanting
– *Traditional*: Seedlings are transplanted to final containers when they reach the second to third true leaf stage, 2 or 3 weeks after sowing.
– *Plugs*: Plants generally remain in plugs for 5–6 weeks prior to transplanting.
Temperature: In stages 3 and 4 of plug production, temperatures are lowered to 17–18°C. After transplanting, 15–17°C temperatures may be maintained. Flowering time, plant height and lateral branching are directly correlated to average temperature (between 10 and 25°C). The warmer the temperature, the faster the flowering, the taller the plant and the fewer branches are produced. Selection of temperature is based on the balance between high quality plants (compact, many branched) and flowering time.
Photoperiod: Petunia are quantitative long day plants, flowering at any photoperiod but flowering more rapidly under long days. The number of lateral branches is also affected by photoperiod. The response to photoperiod is significantly influenced by temperature (thermophotoperiodicity). At temperatures less than 20°C, plants are well branched regardless of photoperiod but flower significantly faster under long days. At temperatures greater than 20°C and short days plants will have more branches than those grown under long days; however, flowering is delayed.
Light: Petunias are high light plants. Low light intensity results in delayed flowering and stretched plants. Low light has the same effect on growth as high temperatures.
Fertilization: At early stages of growth (stage 2 of plug production), 50–75 ppm N of KNO_3 may be applied. While in the plug, nitrogen level may be raised to 150 ppm. Once transplanted and growing actively, plants can receive a balanced application of 150–200 ppm N from a complete fertilizer at each irrigation or 300–400 ppm N once a week. pH should be maintained between 5.5 and 5.8. Boron deficiency is often seen with petunia. This is manifested as hard, distorted and mottled foliage, proliferation of side shoots and terminal bud abortion. Boron deficiency may be avoided by maintaining pH at 5.5–5.8 and supplementing with boron.
Height control: If plants are grown at cool temperatures, height control is not often required. However, if temperatures rise and photoperiod lengthens, then control of height may be necessary. DIF (2–3°C warmer during the night than day) is effective at plug stage. Daminozide (2500–5000 ppm) has been effective when initially applied when plants are about 5 cm across, with one or two additional applications then 7–10 days apart. Paclobutrazol and uniconizol are also effective.

Common problems
Physiological: Most problems with petunias are the result of poor photoperiod/ temperature control. Plants are also sensitive to ozone.
Pests: Relatively trouble free, although aphids can be a persistent problem.
Diseases: Damping off (*Pythium, Rhizoctonia*) and grey mould (*Botrytis*) are the main disease problems.

Postproduction concerns
Fertilization should be reduced prior to shipping. Plants tolerate cold weather and may be displayed in early spring out of doors.

Schedule
Traditional: Plants flower approximately 10 weeks from sowing; however, time is reduced significantly as temperatures and daylength increase in late spring. Baskets require an additional 1–2 weeks.
Plugs: Plants remain in the plug for 5–6 weeks, then an additional 5 weeks is required for flowering in a flat, another week for 10 cm pots.

PRIMULA × *POLYANTHA* (*P. ACAULIS*)

Primrose
Primulaceae

Polyantha primroses or 'polys' are extremely popular bedding plants in areas of mild winters and temperate summers. In northwestern United States, the United Kingdom, northern Europe and New Zealand, polys are one of the leading bedding plants. Throughout the rest of the world, they are popular as pot plants for gifts and home display. Botanically, most commercial cultivars are hybrids of *P. acaulis, P. elatior, P. juliae, P. veris* and other species.

Forms available
Primroses are generally sold in series consisting of many colours. European and Japanese breeders have succeeded in producing large-flowered and miniatures, early and late-flowered and single and bicoloured forms.

Cultivars
Many of the old cultivars were open-pollinated, newer ones are F_1 hybrids.

Crown series, Danova series, Ducat series, Julian mix, Pacific Giant strain, Pageant series, Paloma series, Prominent series and Super Trophy series are some of the cultivars on today's market. All but Pacific Giant strain and Julian mix are F_1 hybrids.

Propagation
All propagated from seed. Seed should be sown on well-drained medium and lightly covered.
Temperature: Temperature of 16–18°C provides optimum germination rate. Germination begins in 7–10 days, after which temperatures may be reduced to 15–17°C (stage 2).
Light: Light is not necessary for germination, but due to non-uniformity of some cultivars, light should be provided so early germinators do not stretch.

Growing-on
Transplanting
– *Traditional*: Transplant 5–7 weeks after sowing to 10 cm pots.
– *Plugs*: Plants remain in plugs for 9–10 weeks prior to transplanting to the final container.
Temperature: Plants should be grown around 15–18°C for the entire crop. Higher temperatures (>20°C) result in faster flowering but quality may not be as good as on plants grown cool. If grown for spring sales, plants of the newer cultivars do not require a cold treatment (<10°C) to initiate flowers (Billingsly and Armitage, 1982). They may be grown more rapidly during the winter if temperatures are not dropped dramatically. When flower buds are visible, temperatures may be reduced to 10°C.
Photoperiod: Plants are quantitative short day plants which flower regardless of photoperiod, but flower faster under short days (<12 hours). Cold temperature is less important for flower initiation when grown under short days compared to long days.
Light: Primroses are best grown in areas of cool temperatures and bright light, particularly if an autumn crop is to be produced.
Fertilization: Plants are not heavy feeders. Stage 2 and 3 plugs may be fed with 25–50 ppm KNO_3 once a week, then applications of 100–125 ppm N from a complete fertilizer are sufficient. If plants are grown cold, reduce application to 75–100 ppm at each irrigation.
Height control: None required.

Common problems
Physiological: Non-uniformity of flowering within a crop is still a problem. 15 to 30 day differences between first plant and last plant flowering are a normal occurrence. At flowering time, only 50% of the bench may be in flower (Armitage, 1992), making for difficulty in sales. This is mainly a problem with the cultivar itself, although treatments (temperature, irrigation, fertilization)

must be uniformly applied. Multiple crowns (more than one growing apex at the same time) are also a function of the cultivar and reduce uniformity and flowering time. Some cultivars produce up to 30% multiple crowns.

Pests: Spider mites, cyclamen mites and thrips can be a problem, but primroses are relatively pest free.

Diseases: Damping off pathogens can be a problem early due to cool, moist conditions in the propagation area.

Postproduction concerns
Plants are best in cool seasons, therefore they are sold relatively early in the spring. They should be protected from frost in the retail area or the garden until they have acclimatized to that site. In areas where summer temperatures are too high for primrose, they are sold as an autumn crop (similar to pansy). It is important that they should not be planted until temperatures in the autumn have moderated.

Schedule
Traditional: Plants require 5–6 months from sowing to flowering in a 10 cm pot.

Plugs: Plants spend 9–10 weeks in the plug, then an additional 10–14 weeks in the pot.

SALVIA SPLENDENS

Salvia
Lamiaceae

The genus contains over 500 species and many are grown as garden plants. Ornamental plant breeders have concentrated their efforts on the red bedding salvia but plants with lavender, pink, white, purple and bicolours are now available in a range of heights.

Forms available
Dwarf (20–30 cm), medium (30–40 cm) and tall (>40 cm) cultivars are available in a range of colours.

Cultivars
Dwarf: 'Fuego' (red), 'Hot Stuff' (red), Sizzler series.
Medium: Carabiniere series, Cleopatra series, 'Pharaoh' (red), 'Red Pillar' (red), Scarlet series, 'St John's Fire' (red), Top series.
Tall: 'America' (red), 'Early Bonfire' (red), 'Rambo' (red).

Propagation
Propagated by seed; cleaned and graded seed are available from seed distributors.
Temperature: Seeds are initially germinated at 24–26°C, then reduced to 21–22°C after about 1 week. Germination is complete around 14 days.
Light: Salvia benefits from light during germination. Seeds should only be lightly covered to maintain high moisture around the seeds. Supplemental light is useful at the earliest stages of germination.

Growing-on
Transplanting
– *Traditional*: Seedlings are transplanted as soon as they can be handled. Transplanting occurs 3–4 weeks after sowing.
– *Plugs*: Seedlings remain for 5–6 weeks in plugs.
Temperature: In plugs, stages 3 and 4 are grown around 15–17°C until transplanting. After transplanting (traditional or plugs), night temperatures of 13–15°C and day temperatures of 21–24°C result in good growth.
Photoperiod: Photoperiod is cultivar dependent. For example, 'America' is short day, 'Carabiniere Red' is day neutral, and 'St John's Fire' is a long day plant. In general, growers do not apply specific photoperiod unless growing for an early or late market.
Light: In the greenhouse, Salvia is a high light plant. Supplemental light while plants are in the seedling stage accelerates growth. Low light levels cause plant stretch.
Carbon dioxide: Young plants of salvia respond well to 1000–1500 ppm CO_2.
Fertilization: Fertilization of 50–75 ppm N of KNO_3 benefits plugs in stages 2 and 3, and once plants are transplanted and actively growing 100–150 ppm N from a complete fertilizer is sufficient. pH should be maintained between 5.5 and 6.2.
Height control: Chemical height control by daminozide (5000 ppm, twice) and chlormequat (750–1000 ppm, two to four times) is effective. DIF is also used for height control.

Common problems
Physiological: Seedlings are highly sensitive to methyl bromide, and will be distorted and stunted if germinated in soil contaminated with minute amounts of the chemical. High soluble salts result in necrotic foliage. Leaf drop is common when plants are stressed.
Pests: Aphids, whiteflies and spider mites are common pests of salvia.

Diseases: *Botrytis* under cool, moist conditions can be a problem. Damping off (*Rhizoctonia* and *Pythium*) are responsible for damping off of seedlings.

Postproduction concerns
Plants should be placed in a shady part of the retail area. They are tolerant of cool weather, although some reddening of foliage may occur when temperatures fall below 10°C.

Schedule
Traditional: Approximately 10 weeks are required from seed to flower, although faster flowering may occur if supplemental light, photoperiod and carbon dioxide are employed.
Plugs: 5 to 6 weeks in the plugs, followed by about 5 weeks in the final container.

SENECIO CINERARIA

Dusty miller
Asteraceae

The grey foliage of dusty miller contrasts well with almost all flower colours. Plants are used in large garden containers or as edgings to complement other bedding plants. The flowers are not important; therefore growers concentrate on growing compact well-branched plants with healthy foliage rather than flowering plants with early or many flowers.

Cultivars
'Cirrus' has shallow indentations while the foliage of 'Silver Dust' and 'Lace' is deeply cut.

Propagation
By seed; some primed seed is available. Due to naturally poor germination, raw seeds are often double-seeded.
Temperature: Germination is most rapid and uniform between 22 and 24°C. Germination should be visible in 4–5 days and be complete in 2 weeks.
Light: Seeds are small and should not be covered. Germination is enhanced by light.

Growing-on

Transplanting
– *Traditional*: Seedlings are transplanted after two to four true leaves have emerged, approximately 3–4 weeks after sowing.
– *Plugs*: Seedlings remain in plug flats 6–7 weeks prior to transplanting to the final container.
Temperature: In stages 3 and 4 of plug growth, temperatures are best maintained between 18 and 21°C (stage 3) and 15 and 16°C (stage 4). After transplanting (traditional or plugs), plants are grown on at night temperatures of 16–18°C, day temperatures around 21°C.
Photoperiod: None.
Light: In the winter and spring, as much light as possible should be provided. In late spring and summer, plants benefit from light shade over the greenhouse. Supplemental light is beneficial in accelerating growth, but is seldom used.
Fertilization: During plug growth, application of 50–75 ppm KNO_3 is recommended. After transplanting, 100–150 ppm N from a complete fertilizer are applied when plants are irrigated.
Height control: Generally not needed, but plants respond to daminozide (two applications of 2500 ppm, 7 days apart) and chlormequat (two or three applications of 1500 ppm, 10 days apart).

Common problems

Pests: Leaf miners are of major concern. They are attracted to dusty miller and do serious damage to the foliage.
Diseases: *Alternaria* is the most important disease.

Postproduction concerns

Although tolerant of cold weather, frost significantly damages the foliage. These plants should not be placed in the outdoor retail area too early.

Schedule

Traditional: Plants are generally ready for sale 10–12 weeks after sowing.
Plugs: Plugs are transplanted 6–7 weeks after sowing, then are grown for an additional 5–7 weeks.

TAGETES ERECTA, T. PATULA

African, French marigold
Asteraceae

Marigolds have long been popular as summer bedding plants and breeders continue to provide new cultivars of African and French forms. Crosses between the two species as well as triploid cultivars have also been offered. Other species of marigolds (*T. tenuifolia*, *T. lucida*) are also occasionally grown as bedding plants.

Forms available
African: African (sometimes called American) marigolds (*T. erecta*), are generally tall (30–50 cm) with double or semi-double flowers.
French: French marigolds (*T. patula*) are most popular with landscapers and gardeners and are generally shorter (<30 cm) and bear single, semi-double or double flowers.
Triploids: Triploids are crosses between the two species and consist of sterile free-flowering plants.
Signet: Signet marigolds (*T. tenuifolia*) are compact plants bearing many single or semi-double flowers over fern-like foliage.

Cultivars
African: Antigua series, 'Cortez', Crush series, Discovery series, Excel series, Inca series, Lady series, Marvel series, 'Monarch', Perfection series, Space Age series.
French: Aurora series, Bonanza series, Boy series, Disco series, 'Golden Gate', Hero series, Janie series, Legend series, Little Hero, Safari series, Sophia series, Spice series.
Triploids: Nugget series, Zenith series.

Propagation
All from seed; de-tailed and coated seeds are available.
Temperature: Seeds are germinated at 24–27°C for the first 3–4 days, then lowered to about 21°C. Germination begins in 2–3 days and should be complete in a week.
Light: Light is not necessary for germination. Seeds may be lightly covered with coarse vermiculite to maintain moisture around the seeds.

Growing-on

Transplanting
– *Traditional*: Seedlings may be transplanted when one or two true leaves have emerged; approximately 1–2 weeks after germination. Seedlings can become quickly overgrown, making transplanting difficult.
– *Plugs*: Seedlings of French marigolds stay in the plug flat for 4–5 weeks; African forms remain for an additional week.
Temperature: Continue stages 3 and 4 of plug growth around 16–18°C. After transplanting, grow plants at 13–16°C night and 18–22°C day temperature. DIF may be used for height control.
Photoperiod: All marigolds are quantitative short day plants. However, the response to short day is displayed more in *T. erecta* and *T. tenuifolia* than in the French marigold, *T. patula*. The critical photoperiod for *T. erecta* is between 12.5 and 13 hours (Tsukamoto *et al.*, 1968) but may differ between species and cultivars (Carlson *et al.*, 1982). Plants of *T. erecta* and *T. tenuifolia* sown in late winter or early spring should be grown under artificial short days for the first 2–3 weeks after germination. Photoperiod control is not as critical with French marigolds and seldom practised for the normal spring crop.
Light: Marigolds are high light plants and stretch under low light levels. Supplemental light is useful in areas of low light to reduce plant stretch.
Fertilization: Seedlings are fertilized with 50–75 ppm N approximately 2 weeks after sowing and raised to 150, then to 200 ppm N constant liquid feed from a complete fertilizer during the growing-on phase. pH is important and should remain between 5.8 and 6.2. Low pH can result in iron toxicity problems. Soluble salts should be monitored and media leached if salt concentration becomes high.
Height control: Usually not a problem in dwarf and semi-dwarf French and Signet types, however some chemicals are effective. Daminozide (2500 ppm, twice, 10 days apart), paclobutrazol (4–10 ppm), ancymidol (50–100 ppm) and chlormequat (750–1500 ppm) are also effective. DIF can be used and reduction of water and fertility is also practised to keep marigolds short.

Common problems

Physiological: Excessive growth in the seed flat or plug tray is common due to the rapid growth of plants. Overgrown seedlings require additional time to become established in the final container.

Low pH releases excess amounts of iron and manganese which may result in brown to black leaf speckling and destruction of the growing tip in susceptible cultivars.
Pests: Spider mites are very common. Aphids, leaf miners and thrips also cause significant headaches. Slugs and snails are a major pest in damp climates.
Diseases: Tomato spotted wilt virus (TSWV), a deadly disease of marigolds and several other bedding crops, is carried by thrips. Damping off (*Rhizoctonia* and

Pythium) and grey mould (*Botrytis*) of flowers can be a problem. Southern bacterial wilt (*Pseudomonas solanacearium*) and bacterial leaf spot (*Pseudomonas syringae* var. *tagetes*) are serious diseases. The former is characterized by stunting, wilting and death and the latter by small black dots on the foliage which turn necrotic. There are no known controls for bacterial diseases and seedlings must be destroyed. Some resistant cultivars are being developed. Clean greenhouses are essential.

Postproduction concerns
Tall cultivars (*T. erecta*) should be sold green. Reducing the temperature of French and Signet marigolds at visible bud enhances shelf-life at the retail level.

Schedule
African marigolds are sown, often in the final container, 5–6 weeks before sale.
Traditional: French and Signet cultivars are ready for sale 9–10 weeks after sowing.
Plugs: Seedlings are transplanted 4–5 weeks after sowing and grown 4–6 weeks in the final container.

VIOLA × *WITTROCKIANA*

Pansy, Viola
Violaceae

Pansies have emerged as one of the most important bedding plant species in recent years. In areas of mild winters, many plants are sold in both the autumn and the spring. Breeders of pansies have developed large and small-diameter bicoloured flowers (faces or blotched), and single-coloured flowers (clear). Nearly all commercial pansies are hybrids, the most dominant species being *V. tricolor*. Violas, also called Johnnie jump-ups, are becoming more popular and have *V. cornuta* as the dominant species. They are generally smaller in stature and bear smaller flowers than pansies.

Cultivars
Pansy: Hundreds are available, most as series.

– *Large-flowered*: Atlas series, Crown series, Imperial series, Majestic Giant series, Medallion series, Rally series, Regal series, Roc series, Super Majestic series, Watercolors series.
– *Small-flowered*: Bingo series, Crystal Bowl series, Delta series, Maxim series, Springtime series, Ultima series, Universal series.
Viola: 'Alpine Summer', Baby series, 'Cuty', 'Helen Mount', Jewel series, 'King Henry', Perfection series, Princess series, Sorbet series.

Propagation

All propagated by seed; primed seeds are available from many companies. They are grown in cold frames, field and greenhouse. The information which follows is for greenhouse culture only.
Temperature: Seeds are often pre-chilled in the seed container or immediately after sowing in the tray for 2 weeks at 5–7°C. After chilling, trays are placed at 17–20°C until emergence. Germination should be complete within 10 days. Temperatures above 24°C significantly inhibit germination.
Light: Light is not necessary for germination.

Growing-on

Transplanting
– *Traditional*: Seedlings are transplanted when they have two or three true leaves.
– *Plugs*: Plugs are grown for 6–7 weeks prior to moving to the final container.
Temperature: Temperature is the most important criterion for high quality pansy production. Cool temperatures are essential. In stage 3 and stage 4 of plug growth, temperatures should be lowered to 15–16°C and 13–15°C respectively. After transplanting, regardless of system used, 13–15°C night temperatures and 16–20°C day temperatures should be maintained. Warmer temperatures result in tall, poorly branched plants. Lower temperatures may be used but flowering time will be delayed.
Photoperiod: Control of photoperiod is not practised with pansies. Studies with *V. tricolor*, an important species of bedding pansies, showed that flowers are produced more rapidly and that leaves are larger under long days than short day conditions (Vince–Prue, 1975).
CO_2: Apply 1000 ppm for 3–5 weeks when plants are young.
Light: Plants are high light plants and require shading only to reduce temperature. Shading is done when cool temperatures cannot be controlled.
Fertilization: Plants are not heavy feeders. Applications of 50 ppm N while in the plug tray enhance growth. Once plants are transplanted, 100–150 ppm N from a balanced fertilizer is used. A pH of 5.5–6.5 is optimal, pH above 6.5 can cause boron deficiency. High soluble salts should be avoided.
Height control: If plants are grown cool (<15°C), height control should not be necessary. For height control, DIF, uniconizol (<10 ppm) and paclobutrazol are effective.

Common problems

Physiological: Warm temperatures result in plant stress, causing a greater incidence of disease and elongated internodes. This becomes critical when plants are grown in the summer for autumn sales.

Pests: Spider mites, thrips and aphids are serious pests.

Diseases: The most serious diseases are caused by *Thielaviopsis*, *Pythium* and *Phytophthora*. All fungi are more prevalent when plants are stressed. Low soluble salts, proper pH management and good air movement help to reduce disease incidence. Powdery mildew can also be a problem.

Postproduction concerns

The success of autumn-marked pansies, at least in the United States, has resulted in some serious problems. Retailers, landscapers and gardeners often put out pansies too early in the late summer or early autumn while soil temperatures are too high. This results in poor root proliferation and spindly top growth. Gardeners and landscapers should be encouraged to wait until cool weather appears before planting in the landscape.

Schedule

Traditional: Pansies are approximately 11–13 week plants, from seed to flower. Growing plants cold results in longer growing time than warm culture.

Plugs: Plugs are generally grown for 6–7 weeks, followed by an additional 5–7 weeks in the final container.

REFERENCES

American Society of Agricultural Engineers (1991) Engineering practices – Commercial greenhouse design and layout. In: *Engineering Standards of the American Society of Agricultural Engineers*, St Joseph, MI, pp. 494–497.

Anon. (1990) Growers: choose your trays. *Greenhouse Grower* 7(11), 77, 80.

Anon. (1991) Get the right plug tray. *Greenhouse Grower* 9(11), 48–49.

Armitage, A.M. (1982) Keeping quality of bedding plant – whose responsibility is it? *Florist Review* 171 (4438), 34–35, 39.

Armitage, A.M. (1983) Determining optimum time of sowing for bedding plants for extended marketing periods. *Acta Horticulturae* 147, 143–152.

Armitage, A.M. (1984) Early fertilization of bedding plant seedlings. *Georgia Commercial Flower Growers Notes* Nov–Dec, 9.

Armitage, A.M. (1985) Supplemental lighting for plugs. *Proceedings of 1985 National Plug Production Conference*, Ames, IA, pp. 60–68.

Armitage, A.M. (1986a) Chlormequat-induced early flowering of hybrid geranium: the influence of gibberellic acid. *HortScience* 21, 116–118.

Armitage, A.M. (1986b) Influence of production practices on post-production life of bedding plants. *Acta Horticulturae* 181, 269–278.

Armitage, A.M. (1988) Supplemental lighting of plugs: basic questions and answers. *Greenhouse Grower* 6(2), 48–49.

Armitage, A.M. (1992) Going down the primrose path. *Greenhouse Grower* 10(5), 65–70.

Armitage, A.M. (1993) *Bedding Plants: Prolonging Shelf Performance*. Ball Publishing, Batavia, IL.

Armitage, A.M. and Billingsley, J.W. (1983) Influence of warm night temperatures on growth and flowering of *Primula × polyantha*. *HortScience* 18, 882–883.

Armitage, A.M. and Kaczperski, M.P. (1992) *Seed-propagated Geraniums and Regal Geraniums*. Timber Press Growers Handbook Series, vol. 1 (rev), Timber Press, Portland, OR.

Armitage, A.M. and Kowalski, T. (1983) Effect of irrigation frequency during greenhouse production on the post production quality of *Petunia hybrida* Vilm. *Journal of the American Society for Horticultural Science* 108(11), 118–121.

Armitage, A.M. and Tsujita, M.J. (1979) The effect of supplemental light source, illumination and quantum flux density on the flowering of seed propagated geranium. *Journal of Horticultural Science* 54(3), 195–198.

Armitage, A.M. and Wetzstein, H.Y (1984) Influence of light intensity on flower initiation and differentiation in hybrid geranium. *Journal of the American Society for Horticultural Science* 19, 114–116.

Armitage, A.M., Carlson, W.H. and Flore, J.A. (1981) The effect of temperature and quantum flux density on the morphology, physiology and flowering of hybrid geraniums. *Journal of the American Society for Horticultural Science* 106, 643–647.

Atwater, B.R. (1980) Germination, dormancy and morphology of the seeds of herbaceous ornamental plants. *Seed Science and Technology* 8, 523–573.

Ball, V. (1987) The exploding world of plugs. *Grower Talks* 50(9), 26–28, 30, 32, 34, 36.

Barrett, J. (1989) Plug height control: choosing the chemical that's right for you. *Greenhouse Grower* 52(10), 46, 48–50.

Barrett, J. and Nell, T.A. (1991) Is Sumagic the answer to your height control problems? *Grower Talks* 54(11), 59–65.

Behe, B.K. and Beckett, L.M. (1991) Season sales summary. *Professional Plant Growers Association News* 24(12), 2–19.

Benson, J. and Kelly, J. (1990) Effect of copper sulfate filters on growth of bedding plants. *HortScience* 25, 1144 (Abst.).

Bewley, J.D. and Black, M. (1985) *Seeds: Physiology of Development and Germination.* Plenum Press, New York.

Bianco, J., Lassechere, S. and Bulard, C. (1984) Gibberellins in dormant embryos of *Pyrus malus* L. cv. Golden Delicious. *Plant Physiology* 116, 185–188.

Biddington, N.L. (1986) The effects of mechanically-induced stress in plants – a review. *Plant Growth Regulation* 4(2), 103–123.

Biddington, N.L. and Dearman, A.S. (1985) The effects of mechanically induced stress on the growth of cauliflower, lettuce, and celery seedlings. *Annals of Botany* 55, 109–119.

Bieloral, H. and Hopkins, P.A.M. (1975) Recovery of leaf water potential, transpiration, and photosynthesis of cotton during irrigation cycles. *Agronomy Journal* 67, 629–632.

Biernbaum, J. (1992) Water and media testing essential to managing the root zone with minimal leaching. *Professional Plant Growers Association Newsletter* 23(6), 2–5.

Black, M. (1980) The role of endogenous hormones in germination and dormancy. *Israel Journal of Botany* 59, 672–676.

Bodger, K. (1985) 4-inch pots for immediate color. In: Mastalerz, J.W. and Holcomb, E.J. (eds), *Bedding Plants III.* Pennsylvania Flower Growers, State College, PA, pp. 502–509.

Borthwick, H.A., Hendricks, S.B., Parker, M.W., Toole, E.M. and Toole, V.K. (1952) A reversible photoreaction controlling seed germination. *Proceedings of the National Academy of Sciences, USA* 38, 662–666.

Borthwick, H.A., Hendricks, S.B., Toole, M.W. and Toole, V.K. (1954) Action of light on lettuce seed germination. *Botanical Gazette* 115, 205–225.

Boyer, J.S. and McPherson, H.G. (1975) Physiology of water deficits in cereal crops. *Advances in Agronomy* 27, 1–23.

Bradford, K.J. (1986) Manipulation of seed water relations via osmotic priming to improve germination under stress conditions. *HortScience* 21, 1105–1112.

Bryer, N. (1967) Modifications de la croissance de la fige de Biyore (*Bryonia dioica*) a la suite d'irritations tactiles. *Compte rendus des seances de l'Academie des Sciences (Paris)* 264, 2114–2117.

Buchanan, R. (1992) Favourite annual cut flowers. In: Proctor, R. (ed.), *Annuals, A Gardener's Guide* Brooklyn Botanic Garden Record 48(4), 45–53.

Bunt, A.C. (1988) *Media and Mixes for Container-grown Plants*. Unwin Hyman, London.

Campbell, L.E., Thimijan, R.W. and Cathey, H.M. (1975) Spectral radiant power of lamps used in horticulture. *Transactions of the American Society of Agricultural Engineers* 18(5), 952–956.

Carlson, W.H. (1976) Production ... back to basics. Temperature and daylength. *Proceedings of the Ninth International Bedding Plant Conference*, Hershey, PA, pp. 205–215.

Carlson, W.H. (1978) Daylength and *Salvia* varieties. *American Vegetable Grower* 26(3), 20–21.

Carlson, W.H. (1992) Should you grow your own plugs or buy them in? *Greenhouse Grower* 10(11), 10–13.

Carlson, W.H. and Heins, R. (1990) Get the plant height you want with graphical tracking. *Grower Talks* 53(9), 62–63, 65, 67–68.

Carlson, W.H. and Johnson, F. (1985) The bedding plant industry, past and present. In: Mastalerz, J.W. and Holcombe, E.J. (eds), *Bedding Plants III*. Pennsylvania Flower Growers, State College, PA, pp. 1–7.

Carlson, W.H., Armitage, A.M. and Mischel, J. (1982) Producing marigolds for profit, a commercial grower's guide. *Michigan State University Extension, Bulletin E-1443*, Michigan State University, East Lansing, MI.

Carlson, W.H., Kaczperski, M.P. and Rowley, E.M. (1992) Bedding plants. In: Larson, R.A. (ed.), *Introduction to Floriculture*, 2nd edn. Academic Press, San Diego, CA, pp. 511–550.

Cathey, H.M. (1954) Chrysanthemum temperature study. C. The effect of night, day and mean temperature upon the flowering of *Chrysanthemum morifolium*. *Proceedings of the American Society for Horticultural Science* 64, 499–502.

Cathey, H.M. (1964) Control of plant growth with light and chemicals. *The Exchange* 141(11), 31–33; 141(12), 33–35.

Cathey, H.M. (1969) Guidelines for the germination of annual pot plant and ornamental herb seeds. *Florist Review* 144(3742), 21–23, 58–60; 144(3743), 18–20, 52–53; 144(3744), 26–28, 75–77.

Clarkson, D.T. and Warner, A.J. (1979) Relationship between root temperature and the transport of ammonium and nitrate ions by Italian and perennial ryegrass (*Lolium multiflorum* and *Lolium perenne*). *Plant Physiology* 64, 557–561.

Cockshull, K.E., Hand, D.W. and Langton, F.A. (1981) The effects of day and night temperature on flower initiation and development in chrysanthemum. *Acta Horticulturae* 25, 101–110.

Craig, R. (1990) Current status of plant breeding and propagation: where are we going in the twenty-first century – a United States perspective. *Acta Horticulturae* 272, 23–32.

Craig, R. and Laughner, L. (1985) Breeding new cultivars. In: Mastalerz, J.W. and Holcomb, E.J. (eds), *Bedding Plants III*. Pennsylvania Flower Growers, State College, PA, pp. 526–539.

Crocker, W. (1936) Effect of the visible spectrum upon the germination of seeds and fruits. In: Duggar, B.M. (ed.) *Biological Effects of Radiation*, vol. 2. McGraw-Hill, New York, pp. 791–828.

Curtice, G.M. and Templeton, A.R. (1987) *Water Quality Reference Guide for Horticulture*. Aquatrols Corporation of America, Pennsaken, NJ.

Daughtrey, M.L. and Horst, R.K. (1990) Biology and management of diseases of greenhouse florist crops. In: *1991 Recommendations for the Integrated Management of Greenhouse Florist Crops. Part II. Management of Pest and Crop Growth*. New York State College of Agriculture and Life Sciences, Cornell University, Ithaca, NY.

Davis, T. (1991) Post-production performance of plugs treated with Bonzi and Sumagic. *Bedding Plants Foundation Research Report* F-044, 1–4.

Dill, R. (1993) Powerhouses of the plug industry. *Greenhouse Grower* 11(11), 10–11.

Eaton, F.M. and Ergle, D.R. (1948) Carbohydrate accumulation in the cotton plant at low moisture levels. *Plant Physiology* 23, 169–187.

Edwards, T.J. (1932) Temperature relations in seed germination. *Quarterly Review of Biology* 7, 428–443.

El-Leboudi, A.E., Ibrahim, I., El-Okeh, A., El-Sayed, A.H. and Taha, M. (1980) Studies on nutritional status and metabolic activities in hardening plants. *Annals of Agricultural Science*, Ain Shams University, Cairo, Egypt, 25(1 and 2), 339–351.

El-Zeftawi, B.M. (1980) Effects of gibberellic acid and cycocel on coloring and sizing of lemon. *Scientia Horticulturae* 12, 177–181.

Enoch, H.Z. (1984) Carbon dioxide uptake efficiency in relation to crop-intercepted solar radiation. *Acta Horticulturae* 162, 137–148.

Erwin, J.E., Heins, R.D. and Karlsson, M.G. (1989) Thermomophogenesis in *Lilium longiflorum*. *American Journal of Botany* 76, 47–52.

Erwin, J.E., Heins, R.D. and Moe, R. (1991) Temperature and photoperiod effects on *Fuchsia* × *hybrida* morphology. *Journal of the American Society for Horticultural Science* 116, 955–960.

EuroFloratech. (1991) Bedding plant production soars. *EuroFloratech* 1(4), 6.

Evenari, M. (1965) Light and seed dormancy. *Encyclopedia of Plant Physiology* 15(2), 804–847.

Evenari, M. and Newman, G. (1952) The germination of lettuce seed II. The influence of fruit coats, seed coat and endosperm upon germination. *Bulletin of the Research Council, Israel* 2, 75–78.

Firth, K.M. (1992) Storing plugs without a cooler. *Greenhouse Grower* 10(11), 54, 56–57.

Flint, L.H. and McAlister, E.D. (1935) Wavelength of radiation in the visible spectrum inhibiting the germination of light-sensitive seed. *Smithsonian Institute Miscellaneous Collection* 94, 1–11.

Flint, L.H. and McAlister, E.D. (1937) Wavelength of radiation in the visible spectrum promoting germination of light-sensitive seed. *Smithsonian Institute Miscellaneous Collection* 96, 1–8.

Fonteno, W.C. (1988a) Know your media, the air, water and container connection. *Grower Talks* 51(11), 110–111.

Fonteno, W.C. (1988b) How to get 273 plugs out of a 273-cell tray. *Grower Talks* 52(8), 68–70, 72, 74, 76.

Fossler, G.M. (1993) First data reported from AFE consumer tracking study. *Illinois State Florist Association Bulletin* 468, 5–8.

Ganmore-Neumann, R. and Kafkafi, U. (1983) The effect of root temperature and NO_3^-/NH_4^+ ratio on strawberry plants. I. Growth, flowering and root development. *Agronomy Journal* 75, 941–947.

Gianfagna, T.J. and Rachmiel, S. (1986) Changes in gibberellin-like substances of peach seed during stratification. *Physiologia Plantara* 66, 154–158.

Goldsberry, K.L. and Holley, W.D. (1962) Carbon dioxide research on roses at Colorado State University. *Colorado Flower Growers Association Bulletin* 187, 3–4.

Graper, D. and Healy, W. (1987) Supplemental lighting benefits bedding plants. *Greenhouse Grower* 5(2), 34–36.

Greenwald, S.M. (1972) Some effects of CCC on the chlorophyll and carotenoid content of jack bean leaves. *American Journal of Botany* 59, 970–971.

Halfacre, R.C., Barden, J.A. and Rollins, H.A. Jr (1969) Effects of Alar on morphology, chlorophyll content and net CO_2 assimilation rate of young apple trees. *Proceedings of the American Society for Horticultural Science* 97, 40–52.

Hamrick, D. (1989) The three cent bedding plant plug. *Grower Talks* 53(8), 44, 46.

Hand, D.W. (1984) Crop responses to winter and summer CO_2 enrichment. *Acta Horticulturae* 162, 45–64.

Hartmann, H.T., Kester, D.E. and Davis, F.T. (1990) *Plant Propagation, Principles and Practices*, 5th edn. Prentice-Hall, New York.

Heins, R.D. and Carlson, W.H. (1990) Understanding and applying graphical tracking. *Greenhouse Grower* 8(5), 73–77.

Heins, R.D. and Erwin, J. (1990) Understanding and applying DIF. *Greenhouse Grower* 8(2), 73–78.

Heins, R.D. and Lange, N. (1992) Development of systems for storage of bedding-plant plugs. *Bedding Plant Foundation Incorporated Research Report* No. F-056.

Heins, R.D. and Wallace, T.F. (1993) Low-temperature storage of alyssum, vinca, New Guinea impatiens and tuberous begonia plugs. *HortScience* 28, 522 (Abst.).

Hendricks, L., Ludolph, D. and Menne A. (1992) Influence of different heating strategies on morphogenesis and flowering of ornaments. *Acta Horticulturae* 305, 9–17.

Herklotz, A. (1964) Influence of constant and daily alternating temperatures on growth and development of *Saintpaulia ionantha* Wendl. *Die Gartenbauwissenschaft* 29, 425–438.

Herner, R.C. (1986) Germination under cold soil conditions. *HortScience* 21(5), 1118–1122.

Heydecker, W. and Coolbear, P. (1977) Seed treatments for an improved performance – survey and attempted prognosis. *Seed Science and Technology* 5, 353–425.

Hicklenton, P.R. (1988) *CO_2 Enrichment in the Greenhouse*. Timber Press Growers Handbook Series, vol. 2. Timber Press, Portland, OR.

Hicklenton, P.R. and Joliffe, P.A. (1978) Effects of greenhouse CO_2 enrichment on the yield and photosynthetic physiology of tomato plants. *Canadian Journal of Plant Science* 58, 801–817.

Hinton-Meade, T.M. (1980) The use of CO_2 enrichment for winter production of spray chrysanthemums. *Annual Review Effort of the Experimental Horticultural Station*, pp. 4–10.

Hofstra, G. and Wukasch, R. (1987) Are you pickling your pansies? *Greenhouse Grower* 5(9), 14–17.

Holcomb, E.J. and Mastalerz, J.W. (1985) Seeding and seedling production. In:

Mastalerz, J.W. and Holcomb, E.J. (eds), *Bedding Plants III*. Pennsylvania Flower Growers, State College, PA, pp. 87–125.

Holley, W.D., Goldsberry, K.L. and Juengling, C. (1964) Effects of CO_2 concentration and temperature on carnations. *Colorado Flower Growers Association Bulletin* 174, 1–5.

Hsiao, T.C. (1973) Plant responses to water stress. *Annual Review of Plant Physiology* 24, 519–570.

Hughes, A.P. and Cockshull, K.E. (1971) The effects of light intensity and carbon dioxide concentraton on the growth of *Chrysanthemum morifolium* cv. Bright Golden Anne. *Annals of Botany* 35, 531–542.

Hughes, A.P. and Freeman, P.R. (1967) Growth analyses using frequent small harvests. *Journal of Applied Ecology* 4, 453–560.

Jaffe, M.J. (1973) Thigmomorphogenesis: The response of plant growth and development to mechanical stimulation with special reference to *Bryonia dioica*. *Planta* 114, 143–157.

Kaczperski, M.P. and Armitage, A.M. (1992) Short-term storage of plug-grown bedding plant seedlings. *HortScience* 27(7), 798–800.

Kaczperski, M.P. and Armitage, A.M. (1993) Preconditioning plug-grown geraniums with temperature and fertility before storage. *HortScience* 28, 572 (Abst.).

Kaczperski, M.P., Carlson, W.H. and Karlsson, M.G. (1991) Growth and development of *Petunia* × *hybrida* as a function of temperature and irradiance. *Journal of American Society for Horticultural Science* 116(2), 232–237.

Kaczperski, M.P., Lewis, P.M. and Armitage, A.M. (1993) CO_2 decreases clogging bench time. *Greenhouse Grower* 11(11), 36, 38, 39.

Kadman-Zahavi, A. and Ephrat, E. (1976) Development of plants in filtered sunlight. II. Effects of spectral composition, light intensity, daylength and red and far-red irradiations on long- and short-day grasses. *Israel Journal of Botany* 25, 11–23.

Kasperbauer, M.J. (1971) Spectral distributon of light in a tobacco canopy and effects of end-of-day light quality on growth and development. *Plant Physiology* 47, 775–778.

Kessler, R. and Armitage, A.M. (1991) Acceleration of *Begonia* × *semperflorens-cultorum* growth using supplemental irradiance. *HortScience* 26(3) 258–260.

Kessler, J.R. and Armitage, A.M. (1993) Effects of carbon dioxide, light and temperature on seedling growth of *Begonia* × *semperflorens-cultorum*. *Journal of Horticultural Science* 68(2), 281–287.

Khademi, M., Koranski, D.S. and Karlovich, P. (1992) Physiology of annual flowering seed. *Bedding Plant Foundation Research Report* No. F-018, 4 pp.

Kimball, B.A. and Mitchell, S.T. (1979) Tomato yields from CO_2-enrichment in unventilated and conventionally ventilated greenhouses. *Journal of the American Society for Horticultural Science* 104, 515–520.

Kirkby, E.A. (1968) Influence of ammonium and nitrate nutrition on the cation–anion balance and nitrogen and carbohydrate metabolism of white mustard plants grown in dilute nutrient solutions. *Soil Science* 105, 133–141.

Koranski, D.S. (1987) Growing plugs from A to Z. *Grower Talks* 50(9), 64–79.

Koranski, D.S. (1988a) Primed seed, a step beyond refined seed. *Grower Talks* 51(9), 24, 26–27, 29.

Koranski, D.S. (1988b) Feed plugs early. *Grower Talks* 51(9), 36.

Koranski, D.S. (1989) Production 101: Sorting the relationship between water quality, feeding programs and media components. *Grower Talks* 53(5), 32, 34, 36, 38, 43.

Koranski, D.S. and Karlovich, P. (1989) Plugs: problems, concerns and recommendations for the grower. *Grower Talks* 53 (8), 28, 30, 32, 34.

Koranski, D.S. and Laffe, S. (1988) Checking out plugs out close . . . *Grower Talks* 52(8), 28–29, 31–32, 34, 36–37, 38, 41–44.

Kramer, P.J. (1950) Effects of wilting on subsequent intake of water by plants. *American Journal of Botany* 37, 280–284.

Krol, A.R. van der, Lenting, P.E., Veenstra, J., van der Meer, I.M., Koes, R.E., Gerats, A.G.M., Mol, J.N.M. and Stuitje, A.R. (1988) An anti-sense chalcone synthase gene in transgenic plants inhibits flower pigmentation. *Nature* 333, 866–869.

Laffe, S. and Styer, R. (1989) Keeping pansies short in the plug flat. *Grower Talks* 53(3), 124, 126, 128.

Lang, A. (1970) Gibberellins: structure and metabolism. *Annual Reviews of Plant Physiology* 21, 537–570.

Lang, G.A., Early, J.D., Martin, G.C. and Darnell, R.L. (1987) Endo-, para-, and ecodormancy. Physiological terminology and classification for dormancy research. *HortScience* 22, 371–377.

Larson, R.A. (1985) Growth regulators in floriculture. *Horticulture Reviews* 7, 399–481.

Latimer, J. (1992) Give brushing a try. *Greenhouse Grower* 10(11), 52–53.

Levitt, J. (1980) *Responses of Plants to Environmental Stresses. Vol. 1. Chilling, Freezing and High Temperature Stresses*, 2nd edn. Academic Press, New York.

Lieberth, J.A. (1991) Electrifying the world. *Greenhouse Grower* 9(11), 10.

Lindstrom, R.S. (1965) Carbon dioxide and its effect on the growth of roses. *Proceedings of the American Society for Horticultural Science* 87, 521–524.

Lockhart, J.A. (1964) Physiological studies on light sensitive stem growth. *Planta* 62, 97–115.

Loomis, W.E. (1925) Studies in transplanting vegetable plants. *Cornell University, Agricultural Experiment Station Memoir* 87, 1–63.

Mastalerz, J.W. (1977) *The Greenhouse Environment. The Effect of Environmental Factors on Greenhouse Crops*. J. Wiley & Sons, New York.

McDonald, M.B. (1980) Assessment of seed quality. *HortScience* 15, 784–788.

McKee, J.M.T. (1981a) Physiological aspects of transplanting vegetables and other crops. I. Factors which influence re-establishment. *Horticultural Abstracts* 51(5), 265–272.

McKee, J.M.T. (1981b) Physiological aspects of transplanting vegetables and other crops. II. Methods used to improve transplant establishment. *Horticultural Abstracts* 51(6), 355–368.

McMahon, M.J., Kelly, J.W., Decoteau, D.R., Young, R.E. and Pollack, R.K. (1991) Growth of *Dendranthemum grandiflorum* (Ramat) Kitamura under various spectral filters. *Journal of the American Society for Horticultural Science* 116, 950–954.

Mechel, B.E. and Kauffmann, A. (1973) The osmotic potential of polyethylene glycol 6000. *Plant Physiology* 51, 914–916.

Metcoff, L. (1992) Nutrition testing: a plant's four basic food groups. *Proceedings of the 1992 International Plug Conference*, Orlando, FL.

Meyer, P., Heidmann, I., Forkmann, G. and Saedler, H. (1987) A new petunia flower color generated by transformation of a mutant with a maize gene. *Nature* 330, 209–235.

Miranda, R.M. and Carlson, W.H. (1980) Effect of timing and number of applications of chlormequat and ancymidol on the growth and flowering of seed geraniums. *Journal of the American Society for Horticultural Science*, 105, 273–277.

Moe, R. (1983) *Campanula*. Report No. 280. Agriculture University of Norway, Department of Horticulture, 18 pp.

Moe, R. (1990) Effect of day and night temperature alternations and of plant growth regulators on stem elongation and flowering of the long-day plant *Campanula isophylla*. *Scientia Horticulturae* 43, 291–305.

Moe, R. and Heins, R.D. (1990) Control of plant morphogenesis and flowering by light quality and temperature. *Acta Horticulturae* 272, 81–89.

Moe, R. and Mortensen, L.M. (1992) Thermomorphogenesis in pot plants. *Acta Horticulturae* 305, 19–25.

Moe, R., Heins, R.D. and Erwin, J.E. (1991) Stem elongation and flowering of the long day plant *Campanula isophylla* Moretti in response to day and night temperature alternations and light quality. *Scientia Horticulturae* 48, 141–151.

Mooreman, G.W. (1985) Diseases. In: Mastalerz, J.W. and Holcomb, E.J. (eds), *Bedding Plants III*. Pennsylvania Flower Growers, State College, PA, pp. 315–326.

Mortensen, L.M. and Moe, R. (1992) Effects of selective screening of the daylight spectrum, and of twilight on plant growth in greenhouses. *Acta Horticulturae* 305, 103–108.

Mortensen, L.M. and Stromme, E. (1987) Effects of light quality on some greenhouse crops. *Scientia Horticulturae* 33, 27–36.

Mumford, F.E. and Jenner, E.L. (1966) Purification and characterization of phytochrome from oat seedlings. *Biochemistry* 10, 98–101.

Nau, J. (1989) *Ball Culture Guide, The Encyclopedia of Seed Germination*. George J. Ball Co., West Chicago, IL, 112 pp.

Nelson, L.J. (1984) *The influence of environmental factors on post production keeping quality of bedding plants*. MS Thesis, Michigan State University.

Nelson, L.J. and Carlson, W.H. (1987) Improve the marketability of bedding plants. *Greenhouse Grower* 5(3), 84–85.

Nelson, L.J., Armitage, A.M. and Carlson, W.H. (1980) Keeping quality of marigolds and impatiens as affected by night temperature and duration. *Florist Review* 167 (4318), 28–29, 62, 74.

Nelson, P.V. (1991) *Greenhouse Operations and Management*. Prentice-Hall, Englewood Cliffs, NJ.

Nickell, L.G. (ed.) (1984) *Plant Growth Regulating Chemicals*, vols I and II. CRC Press, Boca Raton, FL.

Noguchi, H. and Hashimoto, T. (1990) Phytochrome-mediated synthesis of novel growth regulators. A-2a and B, and dwarfism in peas. *Planta* 181, 256–262.

Onofrey, D. (1993) Today's transplanters. *Greenhouse Grower* 11(11), 70, 73–74.

Overy, A. (1992) Annuals with fine foliage. In: Proctor, R. (ed.), *Annuals, A Gardener's Guide*. Brooklyn Botanic Garden Record 48(4), 55–59.

Parups, E.V. and Butler, G. (1982) Comparative growth of chrysanthemum at different night temperatures. *Journal of the American Society for Horticultural Science* 107, 600–604.

Peterson, J.C. (1982) Effect of pH upon nutrient availability in a commercial soilless root medium utilized for floral crop production. *Ohio Agricultural Research and Development Center Research* 268, 16–19.

Peterson, J.C. (1984) Current evaluation ranges for the Ohio State floral crop growing medium analysis program. *Ohio Florists Association Bulletin* 654, 7–8.

Pietsch, G.R. and Carlson, W.H. (1993) Effects of day and night temperature and light level on *Catharanthus roseus*. *HortScience* 28, 502 (Abst.).

Pinchbeck, M. and McAvoy, R.J. (1993) Growth and development of *Catharanthus rosens* under various environmental conditions. *HortScience* 28, 520 (Abst.).

Piringer, A.A. and Borthwick, H.A. (1961) Effects of photoperoid and kind of supplemental light on growth, flowering and stem fasciation of *Celosia*. *American Journal of Botany* 48, 588–592.

Piringer, A.A. and Cathey, H.M. (1960) Effect of photoperiod, kind of supplemental light, and temperature on the growth and flowering of petunia plants. *Proceedings of the American Society for Horticultural Science* 76, 649–660.

Pollock, B.M. and Roos, E.E. (1972) Seed and seedling vigor. In: Kozlowski, T.T. (ed.), *Seed Biology*, vol. 1. Academic Press, New York, pp. 314–387.

Powell, C.C. (1992) Integrated management strategies for the common diseases of bedding plants. *Proceedings of the 1992 International Plug Conference*, Orlando, FL.

Powell, C.C. and Lindquist, R.K. (1992) *Ball Pest and Disease Manual*. Ball Publishing Co., Geneva, IL.

Proctor, R. (ed.) (1992) *Annuals, A Gardener's Guide*. Brooklyn Botanic Garden Record 48(4).

Rajapakse, N.C. and Kelly, J.W. (1992) Regulation of chrysanthemum growth by spectral filters. *Journal of the American Society for Horticultural Science* 117, 481–485.

Raviv, M. (1989) The use of photoselective cladding materials as modifiers of morphogenesis of plants and pathogens. *Acta Horticulturae* 246, 275–284.

Reid, M.S. (1985) Ethylene in the care and handling of ornamentals. *Society of American Florists* June 22–23, 25.

Sawaya, M. (1991) Charge plugs with good nutrition. *Greenhouse Grower* 9(11), 41–42, 44.

Sawaya, M. (1993) Growth regulator guidelines. *Greenhouse Grower* 11(11), 60, 62.

Seeley, J.G. (1985) Finishing bedding plants – effects of environmental factors: temperature, light, carbon dioxide, growth regulators. In: Mastalerz, J.W. and Holcomb, E.J. (eds), *Bedding Plants III*. Pennsylvania Flower Growers, State College, PA, pp. 212–244.

Springer, L. (1992) Annuals for shade. In: Proctor, R. (ed.) *Annuals, a Gardener's Guide* Brooklyn Botanic Garden Record 48(4), pp. 67–71.

Starman, T.W., Kelly, J.W. and Pemberton, H.B. (1989) Characterization of ancymidol effects on growth and pigments of *Helianthus annuus* cultivars. *Journal of the American Society for Horticultural Science* 114, 427–430.

Styer, R.C. (1989) Sow, what's your speciality? *Greenhouse Grower* 7(11), 30–31, 33.

Styer, R.C. and Laffe, S. (1989) How to get your germination rates on the rise. *Grower Talks* 53(8), 55–56, 58, 60.

Tangeras, H. (1979) Modifying effects of ancymidol and gibberelin on temperature induced elongation in *Fuchsia hybrida*. *Acta Horticulturae* 91, 411–417.

Tayama, H.K. and Carver, S.A. (1989) Growth regulator chart. In: Tayama, H.K. (ed.) Floriculture crops insect and mite control, growth regulator, and herbicide booklet, *Ohio Florists Association Bulletin* 711, Columbus, OH, pp. 26–32.

Taylorson, R.B. and Hendricks, S.B. (1977) Dormancy in seeds. *Annual Review of Plant Physiology* 28, 331–354.

Tezuka, T., Sekiya, N. and Ohno, H. (1980) Physiological studies on the action of CCC in Kyoho grapes. *Plant Cell Physiology* 21, 969–977.

Truog, E. (1948) Lime in relation to availability of plant nutrients. *Soil Science* 65, 1–7.

Tsukamoto, Y.H., Imanishi, H. and Yakara, H. (1968) Studies on the flowering of marigold. I. Photoperiodic response and its difference among strains. *Journal of the Japanese Society for Horticultural Science* 37(3), 47–55.

Vanderhoef, L.N., Quail, P.H. and Briggs, W.R. (1979) Red light-inhibited mesocotyl elongation in maize seedlings. *Plant Physiology* 63, 1062–1067.

van Kester, W.N.M. (1990) Current status and perspectives of breeding seed propagated ornamentals. *Acta Horticulturae* 272, 17–21.

Vetanovetz, R.P. and Knauss, J.F. (1988) Water quality. *Greenhouse Grower* 6(12), 64–66, 68–69, 72.

Vince-Prue, D. (1975) *Photoperiodism in Plants*. McGraw-Hill, London.

Vogelezang, J. (1992) Discussion: thermomorphogenesis. *Acta Horticulturae* 305, 63–64.

Vogelezang, J., Cuijpers, L., de Graaf-van der Zande, M.Th. (1992) Growth regulation of bedding plants by reversed day/night temperature only? *Acta Horticulturae* 305, 37–43.

Vollmer, G. (1991) Germination in plugs and plug season 1991. *Proceedings of the 1991 International Plug Symposium*, Chicago, IL.

Wample, R.L. and Culver, E.B. (1983) The influence of paclobutrazol, a new growth regulator, on sunflowers. *Journal of the American Society for Horticultural Science* 108, 122–125.

Warrington, I.J. and Mitchell, K.J. (1976) The influence of blue- and red-biased light spectra on the growth and development of plants. *Agricultural Meteorology* 16, 247–262.

Went, F.W. (1957) *The Experimental Control of Plant Growth*. Chronica Botanica Co., Waltham, MA.

Wilkins, M.B. (1969) *The Physiology of Plant Growth and Development*. McGraw-Hill, London.

Willits, D.H. and Peet, M.M. (1981) CO_2 enrichment in a solar energy collection/storage greenhouse. *American Society of Agricultural Engineers Paper* 81, 15–21.

Wolnick, D.J. and Mastalerz, J.W. (1969) Response of petunia cultivars to selected combinations of electric light, photoperiod, temperature and B-Nine. *Pennsylvania Flower Growers Bulletin* 216, 1–7.

Wright, S.T.C. (1972) Physiological and biochemical responses to wilting and other stress conditions. In: Rees, A.R., Cockshull, K.E., Hand, D.W. and Hurd, R.G. (eds), *Crop Processes in Controlled Environments*. London, Academic Press, Applied Botany Series, vol. 2, pp. 349–361.

Index